5450

Holographic and
Speckle Interferometry

Holographic and Speckle Interferometry

A discussion of the theory, practice and application of the techniques

ROBERT JONES

*Consultant in Optics,
Cambridge Consultants Ltd.,
Cambridge*

CATHERINE WYKES

*Research Fellow,
Department of Mechanical Engineering,
Loughborough University of Technology*

CAMBRIDGE UNIVERSITY PRESS

CAMBRIDGE

LONDON   NEW YORK   NEW ROCHELLE

MELBOURNE   SYDNEY

7142 - 1087

Published by the Press Syndicate of the University of Cambridge
The Pitt Building, Trumpington Street, Cambridge CB2 1RP
32 East 57th Street, New York, NY 10022, USA
296 Beaconsfield Parade, Middle Park, Melbourne 3206, Australia

© Cambridge University Press 1983

First published 1983

Printed in Great Britain by J. W. Arrowsmith Ltd., Bristol BS3 2NT

Library of Congress catalogue card number: 82-1338

ISBN 0 521 23268 6

Contents

Contents   vii

Preface

Holographic and speckle interferometry, which are usually based on laser illumination, enable measurements of displacement (static or dynamic) and shape to be made on optically rough surfaces at sensitivities of the order of the wavelength of light. They can therefore be used to extend the methods of classical optical interferometry to the study of a wide range of objects and systems previously outside the scope of such interferometric investigation. The principle of holographic interferometry was established in the mid-1960s and is based on holographic wavefront reconstruction. Speckle interferometry developed from this work; it relies on the speckle effect which is a random interference pattern observed when coherent light is scattered from a rough surface. In both cases it was the development of lasers capable of generating visible radiation having both high coherence and intensity that enabled the methods to be applied to the solution of practical problems.

Although the techniques are relatively new, their application in such diverse areas as strain and vibration analysis, flow visualization, non-destructive testing and metrology has stimulated a large volume of fundamental and applied research; the results of this work are of considerable importance to a wide range of scientists and engineers. This book provides a self-contained description of the theoretical principles together with a detailed discussion of practical techniques and a survey of applications. The contents may be classified as follows:

Introduction

Chapter 1: This contains an introduction to some basic principles of geometrical optics, diffraction theory, holography and the speckle effect essential to the understanding of the remainder of the book. Although parts of this chapter will not be necessary for readers with some background in optics, specific sections should be studied in accordance with suggestions made in the first section of the chapter.

Basic theory

Chapters 2, 3, 4: These establish the basic principles of Holographic, Speckle and Electronic Speckle Pattern Interferometry (ESPI) respectively with particular reference to displacement measurement and in each case, the factors which limit the application of the technique are discussed.

Shape measurement

Chapter 5: Various methods by which the techniques may be applied to the measurement of surface shape are discussed.

Experimental method

Chapter 6: The purpose of this chapter is to indicate practical aspects which should be taken into account when designing holographic and speckle interferometric systems. It should prove particularly useful to readers who wish to apply the techniques to the solution of their specific problems.

Applications

Chapter 7: This consists of a survey of areas of general application each of which is illustrated by examples of practical problems that have been solved using the various methods.

Appendices

A–C: These summarise important definitions and equations from vector theory, complex number theory and Fourier transform theory respectively which are used in the text.

D–F: These contain derivations and statements of results which it was considered would interfere with the continuity of the main text if included therein.

This monograph will be of particular use to research workers, scientists, and engineers who are either involved in similar work or who wish to become acquainted with the techniques and to apply them to the solution of new problems. It is hoped that in this role it will serve the dual purposes of a standard text and a laboratory manual. Undergraduate and postgraduate students specialising in coherent optics and interferometry should find that the book provides a broad introduction to the subject.

Acknowledgements

The authors have worked in the Laser Optics group at Loughborough University of Technology and would like to thank the head of the group, Professor J. N. Butters, for his continuing help and encouragement. They have also benefited from the opportunity to apply the techniques to practical industrial problems through Loughborough Consultants Ltd. They have had many useful conversations with colleagues at Loughborough, notably Jack Leendertz (who established the principle of speckle correlation interferometry and designed the original Loughborough ESPI system), John Cookson, Mike Smeeton and David Hamson. Many experiments have been made possible by the excellent design and manufacturing work of Vic Roulstone. A large part of the results presented was obtained using equipment and technical support made available by the Department of Mechanical Engineering at Loughborough University. The financial support of the SERC and NRDC is also gratefully acknowledged.

We wish to acknowledge many useful conversations with David Jackson (of Jackson Electronics Ltd) concerning video techniques as applied to ESPI systems.

We would like to thank Dr John Wykes whose comprehensive critique of the first four chapters of the book made a considerable contribution to their final form. We are also very grateful to Jean Jones for her critical review of Chapters 6 and 7.

We are very grateful to Janet Smith who deciphered the often well-nigh illegible handwritten version and converted it into the final typed manuscript, to Jean Geeson who traced the figures and to Ken Topley who prepared the photographic prints.

We would like to thank the following individuals for permission to publish their results:

Dr John Gates (National Physical Laboratory), Figure 2.30 (See Gates, Hall and Ross, Chapter 2, reference 6).

Dr David Rowley (Loughborough University of Technology), Figure 3.10.

Dr Nils Abramson (Royal Institute of Technology, Stockholm), Figures 5.21(*a*), (*b*) and 6.14(*a*) and (*b*).

Don Herbert and David Hamson (Loughborough University of Technology), Figure 7.1.

A. E. Ennos (National Physical Laboratory), Figures 7.8 and 7.9 (See Ennos and Virdee, Chapter 7, reference 9).

Professor John Dent (Loughborough University of Technology), Figure 7.16.

The remaining interferograms were recorded by the authors and included examples of work carried out for the following firms:

BL Cars Ltd
Westland Helicopters Ltd
Noel Penny Turbines Ltd
Leyland Vehicles Ltd

We are also indebted to the above for their permission to reproduce the results.

We would like to thank Lynn Chatterton, Dr Charles Lang, Dr Simon Mitton, Dr Jim Burch and Professor David Whitehouse for their initial encouragement and the staff of CUP for the expert work carried out in the production of the monograph.

Finally we would like to thank our families, in particular Rory, Janet, Jennie and Paul, who bore with us for the many months during which nearly all our spare time was occupied by the writing of this book.

1

Basic optical principles

1.1 Introduction

All the basic optical principles required for the discussion of holographic and speckle interferometry are outlined in this chapter. It is largely self contained and should enable the reader with no background in optics to follow the ensuing discussions. A more detailed treatment will be found in any standard optics textbook – see, for example, references (1)–(5). For those readers who are familiar with basic optics, it may be useful to read Sections 1.4 and 1.9 on Fourier transforms and Sections 1.7 and 1.8 on holography and laser speckle. The discussion in Section 1.8.3 which outlines the approach used throughout the text to calculate phase in coherent imaging is of particular importance since these phase terms form the basis of all the techniques discussed.

1.2 Vibrations and waves

In this section, a basic description of simple harmonic motion and its extension to sinusoidal wave motion is presented. A mathematical description of spherical and plane wavefronts is given, and the concepts of polarization and coherence length introduced.

1.2.1 Simple harmonic motion

A form of oscillation known as simple harmonic motion is fundamental to all kinds of wave motion. A particle is oscillating with simple harmonic motion if its displacement $U(t)$ in a given direction from a fixed point at time t is described by the following equation

$$U(t) = u_0 \sin (2\pi f t + \phi) \tag{1.1}$$

u_0, f, and ϕ are arbitrary constants.

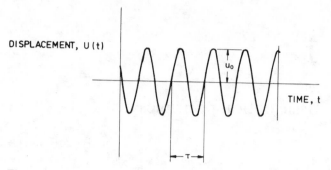

Fig. 1.1 Simple harmonic displacement plotted as a function of time

In Figure 1.1 $U(t)$ is plotted as a function of time. It can be seen that u_0 is the maximum displacement of the particle from the origin and f is the number of times the complete cycle is performed in unit time. The value of ϕ is determined by the position of the particle at $t = 0$, so that its value is determined by the choice of time origin. These quantities are known as:

u_0: amplitude

f: frequency

The argument of the sine function is called the phase of the oscillation and the factor ϕ is the starting phase. We also have $\tau = 1/f$, where τ is the duration of one cycle of the motion, and is known as the period.

An example of an approximation to simple harmonic motion is the oscillation of a body attached to a spring where the amplitude of the oscillation is small. When the body is displaced, it will vibrate about a mean position. An ideal simple harmonic oscillator would continue to oscillate for ever – in the case of the real system, it will gradually slow down due to friction and heat loss.

We see, however, that such a system is initially given energy (by the displacement of the body) and that the system would then retain that energy indefinitely if dissipative forces were not present. The kinetic energy K, of the system is given by

$$K = \tfrac{1}{2}mv^2$$

where v is the velocity of the body.

The velocity obviously changes, and hence the kinetic energy changes. However we may define a potential energy, U, such that the sum of the two remains constant. If U is defined as zero when $K = K_{max}$, then the total energy, E, is given by

$$E = K_{max}$$

When the body passes through the origin, its velocity is at a maximum, having a value of

$$v_{max} = 2\pi f u_0$$

giving the energy of the system as

$$E = 2m\pi^2 f^2 u_0^2 \qquad (1.2)$$

1.2.2 One-dimensional wave motion

We consider first a wave being propagated along a string since this can be readily visualized. Imagine one end of the string to be vibrated up and down so that its motion is described by equation (1.1). A point adjacent to the end of the string will be forced to oscillate in the same way, but will be slightly behind the first point in its motion. Each subsequent point on the string will be oscillating with simple harmonic motion, but with its phase retarded with respect to previous points. The motion of a point which is located at a distance z along the z-axis is therefore described by the equation:

$$U(z, t) = u_0 \sin [2\pi ft - \phi(z)] \qquad (1.3)$$

If we assume that the phase factor $\phi(z)$ is linearly proportional to the distance z from the origin, equation (1.3) can be written as

$$U(z, t) = u_0 \sin (2\pi ft - kz + \phi) \qquad (1.4)$$

where k is a constant.

In Figure 1.2, $U(z, t)$ is shown as a function of z for a time t_1 and a later time t_2 where $(t_2 - t_1) < \tau$. It is seen that a sinusoidal waveform travels along the string. The distance between peaks is given by $2\pi/k$, which is generally written as

$$\lambda = 2\pi/k \qquad (1.5)$$

and is known as the wavelength.

A given part of the wave will travel through one wavelength during one cycle of the vibration. Thus, the speed at which the waveform travels is given by:

$$c = \lambda f \qquad (1.6)$$

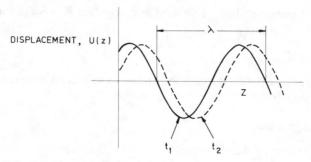

Fig. 1.2 A sinusoidal waveform plotted at times t_1 and t_2

It is important to realize that there is no movement of the string in the z-direction; the individual points move up and down and only the waveform travels in the z-direction.

However, since the motion of points along the string is caused by the motion of the end of the string, it is clear that energy must be supplied to that point in order to maintain the motion.

The rate at which energy passes a given point in unit time is known as the intensity of the wave. The following derivation shows how the intensity of the wave is related to the amplitude and frequency of the wave.

The string is considered to be made up of a series of particles of mass m; there are N particles/unit length of the oscillating string and the total energy E_λ contained in a section of the string one wavelength long is given from equation (1.2) by

$$E_\lambda = 2mN\lambda\pi^2 f^2 u_0^2 \tag{1.7}$$

During one cycle of the motion, an amount of energy E passes a given point so that the amount of energy which passes a given point in unit time is given by

$$I = \frac{E_\lambda}{\tau} = \frac{cE_\lambda}{\lambda} = 2mNc\pi^2 f^2 u_0^2 \tag{1.8}$$

Thus, the intensity of the wave is proportional to the square of its amplitude.

The intensity of a two-dimensional wave is defined as the rate at which energy passes through a line of unit length. For a three-dimensional wave, the intensity is the rate at which energy passes through unit area.

It can be shown in both these cases that the intensity is again proportional to the square of the amplitude.

1.2.3 *Circular and spherical wavefronts*

In the previous section, a one-dimensional wave travelling along a string was discussed. A waveform travelling in two dimensions which is familiar to everyone is a water wave; if a stone is dropped into a still pool, a circular wave propagates from the point at which the stone was dropped; this wave will of course die away after a short while. If, however, we imagine a stick being oscillated in simple harmonic motion at a point in the water, a circular wave will spread out from that point, transmitting energy as it does so. The size of the wavefront increases as the square of the distance, r, from the origin of the wave and the energy is distributed over an increasingly large area which increases as r. Without going into the details of surface gravity waves, it can be assumed that the intensity of the wavefront decreases as $1/r$.

A spherical wave propagates uniformly from a fixed point; the area covered by the wavefront increases as r^2, so that the intensity will decrease as $1/r^2$.

1.2.4 *Plane wavefronts*

A plane wavefront is one in which the phase at all points in a plane perpendicular to the direction of propagation of the wave motion is constant. A wavefront is transmitted in the z-direction whose amplitude is described by

$$U(z, t) = u_0 \sin (2\pi ft - kz) \text{ for all } x \tag{1.9}$$

– see Figure 1.3.

The displacement described by equation (1.9) may be in any direction. In the case of a sound wave it is the direction of propagation; in the case of an electromagnetic wave, the disturbance, which is a variation in electric and magnetic fields is perpendicular to the direction of motion of the wave.

This is the same equation as equation (1.4) for the wave transmitted along the string, with $\phi = 0$. The plane wave behaves like a set of one-dimensional waves which are mutually parallel. The wavefront then covers the same area in space indefinitely. When the individual one-dimensional waves are not mutually parallel, the wavefront diverges or converges.

A plane wavefront which is propagated in a general direction is described by the equation

$$U(x, y, z, t) = u_0 \sin [2\pi ft - k(n_x x + n_y y + n_z z)] \tag{1.10}$$

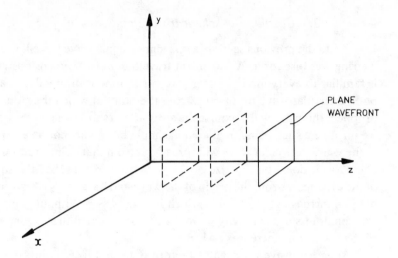

Fig. 1.3 A plane wavefront propagating in the z-direction

where n_x, n_y and n_z are the cosines of the angles between the direction of propagation and the x-, y- and z-axes respectively.

Equation (1.10) may be written in vector notation – see Appendix B – as

$$U(\mathbf{r}, t) = u_0 \sin [2\pi ft - k\mathbf{n} \cdot \mathbf{r}] \qquad (1.11)$$

where

$$\mathbf{r} = x\mathbf{i} + y\mathbf{j} + z\mathbf{k}$$

and

$$\mathbf{n} = n_x\mathbf{i} + n_y\mathbf{j} + n_z\mathbf{k}$$

Equation (1.11) is often written in the form

$$U(\mathbf{r}, t) = u_0 \sin [2\pi ft - \mathbf{k} \cdot \mathbf{r}] \qquad (1.12)$$

where

$$\mathbf{k} = (2\pi/\lambda)\mathbf{n}$$

1.2.5 *Electromagnetic waves*

Whenever an electric charge is accelerated, an electromagnetic wave radiates from the charge at a velocity of $\sim 3 \times 10^8$ m/s (6, 7). The wave is observed as a fluctuating electric field and an associated magnetic field; these are orthogonal to the direction of propagation of the wave and also to each other. The form in which the wave is detected (radio wave, visible light, X-ray, γ-ray etc) depends on the frequency of the oscillating fields. Here we are concerned with visible electromagnetic radiation which has a frequency of $\sim 10^{15}$ Hz, and in particular with the radiation produced by a laser which approximates quite closely to a plane wavefront (the divergence of a laser beam is typically $\sim 10^{-3}$ radians).

In the discussion here, it is necessary to consider only one of the displacements associated with the electromagnetic wave; this will be taken as the electric vector. Thus, the displacement $U(r, t)$ represents the variation of the electric field in the propagating light wave.

The shape of a laser wavefront can be changed by the use of conventional optical components; for example a simple lens converts the plane wavefront laser output into a diverging spherical wavefront – see Figure 1.14.

1.2.6 *Polarization*

The wave motion described by equation (1.3) describes an oscillation in a direction perpendicular to the z-direction, which is being propagated in the z-direction. In the case of a light wave, the oscillation is a variation in the electric field, and it can lie in any direction perpendicular to the z-direction. The direction of this oscillation is defined as the direction of polarization of the light.

In the case of an ordinary light source (sunlight, light bulb, etc) the direction of polarization varies randomly. The output beam of a laser is usually polarized in the vertical direction when the output beam travels horizontally. When a wavefront is polarized in a single direction such as this, it is said to be plane polarized.

Any light wave may be described as a combination of two orthogonal polarization components. Consider for example a light wave of amplitude u_0 travelling in the z-direction which is polarized in a direction forming an angle θ with the x-axis; the amplitude of the light in the x- and y-directions is given by $u_0 \cos \theta$ and $u_0 \sin \theta$ respectively.

The polarization of a wave may rotate as the wave propagates. This is known as elliptical polarization, and can be described by a wave with amplitudes in the x- and y-directions given by

$$U_x = u_{x0} \sin (2\pi ft + \phi_1) \tag{1.13a}$$

$$U_y = u_{y0} \sin (2\pi ft + \phi_2) \tag{1.13b}$$

where $\phi_1 \neq \phi_2$.

The polarization direction describes an ellipse of semi-axes u_{x0}, u_{y0} as the wave propagates.

When $u_{x0} = u_{y0}$ and $|\phi_1 - \phi_2| = \pi$, the polarization direction describes a circle as it propagates. Such a wave is 'circularly polarized'.

Devices known as polarizers are available which will select one direction of polarization only from an input beam, or which will convert plane polarized light into circularly or elliptically polarized light.

1.2.7 Coherence length

The plane wavefront described by equation (1.9) extends to infinity both in time and in space and it has a single frequency which does not change. No real light source has these properties. A laser beam will, however, approximate to a sine wave of fixed frequency over a distance which is specific to a given laser. This distance is known as the 'coherence length' of the laser. With this coherence length is associated a coherence time, i.e. the time during which the wave maintains its sinusoidal form.

Coherence will be discussed more fully in Section 1.4.2.

1.3 The superposition of waves

In this section, the principle of superposition of waves is discussed. The use of complex notation to describe a light wave is introduced and the application of the superposition principle to two-beam interference and to Fourier transform theory is described.

1.3.1 The principle of superposition and interference of waves

The principle of superposition of waves states that if two or more wavefronts are travelling past a given point, the total amplitude of the displacement at that point is given by the sum of the individual displacements; a system which obeys this principle is a linear system. When very high intensity light sources are used it cannot be applied and non-linear optics must be employed. However, all the topics discussed in this book fall within the field of linear optics. If the waves are polarized in the same direction, the amplitudes are added algebraically; if the polarizations are different, then a vector addition is required. (This was implicitly assumed in the discussion of elliptically polarized light in Section 1.2.5). It will be assumed henceforth that all waves are polarized in the same direction, unless otherwise specified.

Consider two wavefronts incident at a point given by

$$U_1 = u_0 \sin (2\pi ft + \phi_1)$$

$$U_2 = u_0 \sin (2\pi ft + \phi_2)$$

The total amplitude is given by

$$U_T = U_1 + U_2 = 2u_0 \sin \left(2\pi ft + \frac{\phi_1 + \phi_2}{2}\right) \cos \left(\frac{\phi_1 - \phi_2}{2}\right) \quad (1.14)$$

The resultant motion is a simple harmonic oscillation of frequency f, and amplitude

$$u_T = 2u_0 \cos \left(\frac{\phi_1 - \phi_2}{2}\right) \quad (1.15)$$

Thus when the phases of the two wavefronts are the same, the amplitude is the sum of the incident amplitudes, but when $(\phi_1 - \phi_2) = \pi$, the amplitude is zero. In the former case, the wavefronts are said to be in phase, and in the latter they are in anti-phase.

Since the intensity of the wavefronts is proportional to the square of the amplitude, the total intensity I_T is given by

$$I_T \propto 4u_0^2 \cos^2 \left(\frac{\phi_1 - \phi_2}{2}\right)$$

which may be written as

$$I_T = 4I_0 \cos^2 \left(\frac{\phi_1 - \phi_2}{2}\right) \quad (1.16)$$

where I_0 is the intensity due to a single wave (see equation 1.7). Thus, the intensity of the sum of the waves may be twice the intensity of the sum of the individual intensities, may be zero, or may have any intermediate value. This effect is known as interference. It is extremely important and is the basis of nearly all the topics and techniques discussed in this book.

Light waves have a frequency of $\sim 10^{15}$ Hz. Currently available detectors do not have this frequency resolution so that a detector which gives an output which is proportional to the amplitude of the input light will average the amplitude over many cycles, giving a resultant amplitude of zero. However, a detector whose output is proportional to the square of the amplitude, for example, photographic emulsions and photo-electric devices, will give a non-zero output. For example, an input signal of amplitude U given by

$$U = u_0 \sin (2\pi f t + \phi)$$

will give an output signal S given by

$$S = \frac{1}{T} \int_0^T U^2 \, \mathrm{d}t = \frac{u_0{}^2}{2}$$

where T covers many periods of the oscillation. Thus, the output signal is proportional to the intensity of the light. We see, however, that no information about the phase is obtained.

Consider now equation (1.16); when two waves at the same frequency but different phases are added together, the resultant intensity depends on their relative phase, so that in this case, an intensity detector does indicate phase information. It is the ability of interference to give phase as well as intensity information which is the basis of its importance in optical measurement techniques.

1.3.2 *The complex representation of a light wave*

The solution of equation (1.14), to give the results (1.15) and (1.16) for the displacement and intensity of the sum of two waves was fairly simple, because the amplitudes of the two waves were the same. If, however, the amplitudes are different, the analysis is more tedious and when three or more waves are involved, the calculation becomes very lengthy. When the light displacement is represented by a complex number, the computation is considerably simplified. (The properties of complex numbers required here are summarized in Appendix A.)

In this representation, the light displacement is described by a complex number

$$U(r, t) = u_0 \exp \mathrm{i}(2\pi f t + \phi) \tag{1.17}$$

and $U(r, t)$ is known as the complex amplitude of the light; u_0, f and ϕ correspond to the equivalent terms in equation (1.1).

Since the intensity is proportional to $u_0{}^2$, we have

$$I \propto UU^* \tag{1.18}$$

Now, consider the light waves represented by complex amplitudes U_1, U_2 given by

$$U_1 = u_1 \exp \mathrm{i}(2\pi f t + \phi_1)$$

$$U_2 = u_2 \exp \mathrm{i}(2\pi f t + \phi_2)$$

The total complex amplitude is given by

$$U_T = u_1 \exp i(2\pi ft + \phi_1) + u_2 \exp i(2\pi ft + \phi_2) \tag{1.19}$$

This may be written as

$$U_T = (u_1 \exp i\phi_1 + u_2 \exp i\phi_2) \exp i2\pi ft \tag{1.20}$$

The intensity is then readily found from

$$I \propto U_T U_T^* = u_1{}^2 + u_2{}^2 + u_1 u_2 [\exp i(\phi_1 - \phi_2) + \exp -i(\phi_1 - \phi_2)] \tag{1.21}$$

Using the relationship

$$(\exp ix + \exp -ix) = 2 \cos x$$

equation (1.21) becomes

$$U_T U_T^* = u_1{}^2 + u_2{}^2 + 2u_1 u_2 \cos (\phi_1 - \phi_2).$$

Hence

$$I_T = I_1 + I_2 + 2\sqrt{I_1 I_2} \cos (\phi_1 - \phi_2) \tag{1.22}$$

We see from equation (1.21) that when the intensity is calculated, the time dependence of the waves disappears; this is the main reason why the complex notation is useful.

1.3.3 *Superposition of waves of different frequencies*

Consider two wavefronts of different frequencies given by

$$U_1 = u_1 \exp i(2\pi f_1 t + \phi_1)$$
$$U_2 = u_2 \exp i(2\pi f_2 t + \phi_2)$$

When these are superimposed the total complex amplitude is

$$U = u_1 \exp i(2\pi f_1 t + \phi_1) + u_2 \exp i(2\pi f_2 + \phi_2)$$

The instantaneous value of the intensity is then given by

$$I \propto UU^* = u_1{}^2 + u_2{}^2 + 2u_1 u_2 \cos [2\pi (f_1 - f_2)t + (\phi_1 - \phi_2)] \tag{1.23}$$

We see that the intensity varies with time. If the frequency difference $\Delta f = f_1 - f_2$ is greater than the frequency resolution of the detector, the last term averages to zero and the mean intensity becomes

$$\langle I \rangle = I_1 + I_2$$

If however, Δf is within the frequency range of the detector, the output intensity with time between a minimum value of $I_1 + I_2 - 2\sqrt{I_1 I_2}$ and a maximum value of $I_1 + I_2 + 2\sqrt{I_1 I_2}$. The phase of this oscillation is determined by $(\phi_1 - \phi_2)$, so that measurement of the phase of the intensity variation enables the phase difference between the two waves to be determined. It is sometimes stated that interference cannot be obtained between waves of different frequencies, implying that phase information cannot be obtained by superimposing two such waves. We see that this is not correct; indeed this approach forms the basis of a high resolution holographic measurement technique – see Section 2.7.1. However, until the advent of lasers and suitable frequency shifting devices, such effects could not be observed.

1.3.4 *Two-beam interference*

Two plane wavefronts have complex amplitudes described from equations (1.12) and (1.17) by

$$U_1 = u_1 \exp \mathrm{i}(2\pi ft - \boldsymbol{k}_1 \cdot \boldsymbol{r})$$

$$U_2 = u_2 \exp \mathrm{i}(2\pi ft - \boldsymbol{k}_2 \cdot \boldsymbol{r})$$

where $\boldsymbol{k}_{1,2} = (2\pi/\lambda)\boldsymbol{n}_{1,2}$.

Omitting the time dependence (Section 1.3.2), we have

$$U_1 = u_1 \exp(-\mathrm{i}\boldsymbol{k}_1 \cdot \boldsymbol{r})$$

$$U_2 = u_2 \exp(-\mathrm{i}\boldsymbol{k}_2 \cdot \boldsymbol{r})$$

The intensity at a point $\boldsymbol{r}$ is given by

$$I = I_1 + I_2 + 2\sqrt{I_1 I_2} \cos(\boldsymbol{k}_1 - \boldsymbol{k}_2) \cdot \boldsymbol{r} \qquad (1.24)$$

For simplicity, the axes are chosen so that $\boldsymbol{n}_1$ and $\boldsymbol{n}_2$ lie in the xz-plane and make equal but opposite angles θ with the z-axis (see Figure 1.4). Hence

$$\boldsymbol{n}_1 = \sin\theta\,\boldsymbol{i} + \cos\theta\,\boldsymbol{k}$$

$$\boldsymbol{n}_2 = -\sin\theta\,\boldsymbol{i} + \cos\theta\,\boldsymbol{k}$$

and

$$(\boldsymbol{k}_1 - \boldsymbol{k}_2) = (4\pi/\lambda)\sin\theta\,\boldsymbol{i}.$$

Consider the time-averaged light intensity along a line parallel to the x-axis. Such a line is described by the vector $\boldsymbol{r} = x\boldsymbol{i} + z_0\boldsymbol{k}$ where z_0 is

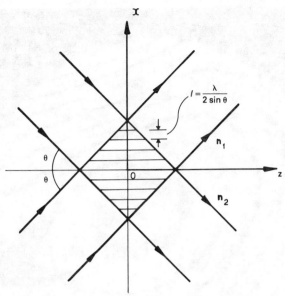

Fig. 1.4 The interference of two plane wavefronts propagating at equal angles to z-axis

the position of the line on the z-axis. Equation (1.24) becomes

$$I = I_1 + I_2 + 2\sqrt{I_1 I_2} \cos \left(\frac{4\pi}{\lambda} x \sin \theta \right) \qquad (1.25)$$

Hence, the intensity varies sinusoidally along any line parallel to the x-axis. The distance L between equivalent points in the intensity variation is given by

$$L = \frac{\lambda}{2 \sin \theta} \qquad (1.26)$$

This intensity variation is known as an interference fringe pattern. It will exist as long as the beams overlap. The maxima and minima of the light intensity are located at constant distances from the z-axis.

Thus, when two plane wavefronts of intensities I_1 and I_2 are superimposed, the intensity varies sinusoidally between a maximum value $(I_1 + I_2 + 2\sqrt{I_1 I_2})$ and a minimum value $(I_1 + I_2 - 2\sqrt{I_1 I_2})$; this intensity variation is known as a fringe pattern. The fringe pattern is in the form of a series of planes of uniform intensity which are parallel to the plane that bisects the angle between the two beams and is perpendicular to the plane defined by the two propagation vectors.

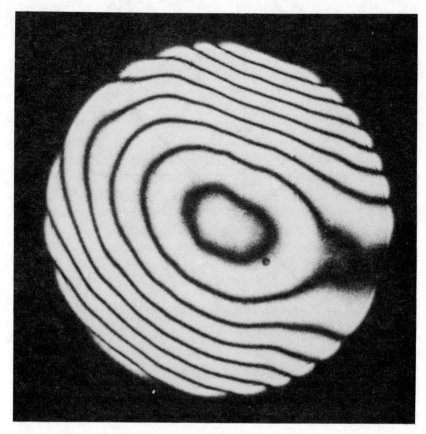

Fig. 1.5 An interference fringe pattern

When two spherical wavefronts are superimposed, an interference pattern is obtained which consists of a set of hyperboloids whose foci are the centres of the wavefronts – an example is shown in Figure 1.5.

Any pair of wavefronts of the same frequency which are added together will give rise to interference. The shape and the spacing of the fringes obtained depend on the nature of the wavefronts and may not be observable; for example, the fringe spacing may be too fine to be resolved: if the two wavefronts each contain many frequencies, each pair of frequencies will have its own interference pattern and these will in general vary so much that the interference effects may not be resolvable.

To observe interference effects, the light must be monochromatic, (i.e. of a single frequency) or very nearly so. This is why interference experiments are easier to perform with laser light than with conventional light sources.

Finally, it should be noted that when two orthogonally polarized waves are added together, the resultant wave motion is a vector which rotates – see Section 1.2.6. The amplitude U_r of this rotating vector is related to the amplitudes of the individual waves U_{x0}, U_{y0} by

$$U_r^2 = U_{x0}^2 + U_{y0}^2;$$

it is seen that it is independent of the relative phases of the two waves, so that interference effects are not observed when orthogonally polarized waves are added together. This must always be borne in mind when interference experiments are being performed.

1.4 Fourier transform theory

Fourier transform theory is a very important mathematical tool in the discussion of waves. A detailed description is to be found in references (8), (9) and (10). In Appendix C, Fourier transforms are defined and some of the theorems which apply to them are given. In this section, the application of Fourier transforms to light waves is discussed.

Any well-behaved function $g(x)$ can be represented by an equation of the form

$$g(x) = \int_{-\infty}^{+\infty} G(f) \exp(2\pi i f x)\, df \qquad (1.27)$$

Thus, the function $g(x)$ is represented by a continuous sum of sine waves of frequency f with amplitudes $G(f)$; the function $G(f)$ is a frequency distribution.

A two-dimensional function $g(x, y)$ may be written as

$$g(x, y) = \int\!\!\int_{-\infty}^{+\infty} G(f_x, f_y) \exp[2\pi i(f_x x + f_y y)]\, df_x\, df_y \qquad (1.28)$$

$G(f)$, $G(f_x, f_y)$ are known as the Fourier transforms of the functions $g(x)$ and $g(x, y)$.

The decomposition of a disturbance of arbitrary structure into a set of simple sinusoidal wave forms simplifies the solution of innumerable problems in physics and engineering; Fourier analysis is used in fields as diverse as electric circuitry, acoustics, scattering of X-rays and electron beams and, of course, optical imaging and holography. Because the response of a system to a simple sinusoidal input is much more readily found than the response to an arbitrary input, the response to the latter is found by summing the responses to the individual Fourier components of the input.

1.4.1 *Fourier representation of light fields of complicated form*

The function $U(\boldsymbol{r})$ which describes the complex amplitude of the light scattered by a surface to a point $\boldsymbol{r}$ is in general a complicated function. (A detailed discussion of this topic is given in Section 1.8.) In many of the systems discussed in this book, the behaviour of such a light field is best understood by expressing it as the sum of a set of plane waves.

The wavefront $U(\boldsymbol{r})$, which is monochromatic, is incident on the $z = 0$ plane, and its amplitude there is given by $U(x, y, 0)$. From equation (1.28), it may be written as

$$U(x, y, 0) = \int\limits_{-\infty}^{+\infty}\!\!\int u(f_x, f_y) \exp\left[2\pi i(f_x x + f_y y)\right] df_x \, df_y \qquad (1.29)$$

In Section 1.2.4, we saw that a plane wavefront travelling in a direction $\boldsymbol{k}$ has a complex amplitude

$$U_p(\boldsymbol{r}) = u_0 \exp \frac{2\pi i}{\lambda}(n_x x + n_y y + n_z z) \qquad (1.30)$$

Thus, we see that the Fourier components of equation (1.29) represent plane waves travelling in directions of direction cosines

$$n_x = \lambda f_x$$
$$n_y = \lambda f_y \qquad\qquad (1.31)$$
$$n_z = \sqrt{1 - n_x^2 - n_y^2}$$

Thus, the light field in the $z = 0$ plane is expressed as the sum of a set of plane waves travelling in different directions and of varying amplitudes; f_x, f_y are known as the spatial frequencies of the waves.

If the light field $U(\boldsymbol{r})$ is obtained by illuminating a transparency of varying transmission described by a function $\rho(x, y)$, with a plane wavefront of amplitude U_0, giving a transmitted amplitude of

$$U(\boldsymbol{r}) = U_0\rho(x, y),$$

then the Fourier components of $U(r)$ are given by

$$A(f_x, f_y) = \iint U_0 \rho(x, y) \exp[-2\pi i(f_x x + f_y y)]\, dx\, dy \qquad (1.32)$$

so that the Fourier components of the wavefront $U(r)$ represent the Fourier components of the function $\rho(x, y)$.

A similar expression is obtained in the case of a wavefront scattered from a surface of varying reflectivity.

It will be seen in Section 1.9 that it is possible to perform spatial frequency filtering on such a wavefront by using a lens and an aperture(s) suitably located.

1.4.2 *Multifrequency light waves and the characteristics of laser light*

The variation with time at a point P of the complex amplitude $U(t)$ of a light wave which is not a sine wave of a single frequency can be written as a sum of frequencies as

$$U_P(t) = \int_0^\infty g(f) \exp(2\pi i f t)\, df \qquad (1.33)$$

where $g(f)$ is the Fourier transform of $U(t)$. We write

$$g(f) = a(f) \exp i\phi(f) \qquad (1.34)$$

where $a(f)$ is real, and represents the amplitude of a given frequency component and $\phi(f)$ represents the phase of that component at $t = 0$. ($U(t)$ can be expressed as an integral between 0 and ∞, instead of between $-\infty$ and ∞ by suitable choice of $\phi(f)$.)

Equation (1.33) can be used to represent the light disturbance from any light source, e.g. sunlight, light bulb, sodium lamp. The output of a laser can however be represented more simply as

$$U(t) = \sum_{p=-n}^{+n} g_p \exp 2\pi i(f_0 + p\Delta f)t \qquad (1.35)$$

where $g_p = a_p \exp i\phi_p$ because it can be considered to consist of a set of sine waves known as modes, separated in frequency by intervals Δf centred on the frequency f_0. These modes arise from the structure of the laser which consists of a resonant cavity formed by two mirrors containing a suitably excited lasing medium (see for example Table 6.2). The interval Δf is related to the length L of the cavity by $\Delta f = c/2L$ where c is the velocity of light. (A detailed discussion of the operation of lasers is outside the scope of this book since we are concerned here with the properties of laser light rather than its generation. References (11) and (12) give a description of the mode of operation of various forms of laser.)

Equation (1.35) suggests that the individual modes extend to infinity in time in both directions. Clearly this is not possible; however, the waves do exist for times typically of the order of 10^{-7} s corresponding to distances of 30 m and when special precautions are taken, they may exist for times of ~ 1 s corresponding to distances $\sim 10^8$ m. Since 10^8 or more cycles will occur during a time of 10^{-7} s, the wave behaves as a sine wave within this time. This is equivalent to saying that we treat the individual components of equation (1.35) as being single frequencies even though strictly speaking a wave which exists only for a finite time has a finite spread of frequencies.

Equation (1.35) can be rewritten as

$$U(t) = \exp 2\pi i f_0 t \sum g_p \exp (2\pi i p \Delta f t) \tag{1.36}$$

and we see that g_p represents the Fourier transform of a periodic function at frequency Δf (see equation (C.1), Appendix C). This equation describes a sine wave of frequency f_0, whose phase and amplitude are modulated by the periodic function

$$M(t) = \sum g_p \exp (2\pi i p \Delta f t) \tag{1.37}$$

$M(t)$ can be seen from equation (1.36) to be a periodic function of frequency Δf, since it is the sum of a set of frequencies which are multiples of Δf.

The form of the modulation function depends on the number of modes present and their relative phases and amplitudes. For example, when only two modes of equal amplitude are present, the modulation function is simply:

$$M(t) = \cos (2\pi \Delta f t)$$

We will now consider what happens when a light wave of the form of equation (1.35) is used to give two beam interference. The light is divided into two beams, one of which travels a distance $c\tau$ further than the other before they are superimposed at an angle to one another. The amplitude can be described by

$$U_T = U_1 M(t) \exp [2\pi i f_0 t - i\mathbf{k}_1 \cdot \mathbf{r}]$$
$$+ U_1 M(t+\tau) \exp [2\pi i f_0(t+\tau) - i\mathbf{k}_2 \cdot \mathbf{r}] \tag{1.38}$$

To find the intensity, we first write

$$U_T U_T^* = U_1^2 \llbracket M(t)M^*(t) + M(t+\tau)M^*(t+\tau)$$
$$+ 2\,\mathrm{Re}\,\{M(t)M^*(t+\tau)\exp\left[-2\pi i f_0\tau - i(k_1 - k_2)\cdot r\right]\}\rrbracket$$

$$(1.39)$$

The intensity is proportional to $U_T U_T^*$ and

$$I = 2I_0 \llbracket M(t)M^*(t)$$
$$+ \mathrm{Re}\,\{M(t)M^*(t+\tau)\exp i[-2\pi f_0\tau - (k_1 - k_2)\cdot r]\}\rrbracket \quad (1.40)$$

The spatial intensity variation in the overlap region due to the term $\exp i[-2\pi f_0\tau - (k_1 - k_2)\cdot r]$ gives rise to an interference pattern. The visibility of an interference pattern (which is a measure of the contrast of the fringes) is defined as

$$V = \frac{I_{\max} - I_{\min}}{I_{\max} + I_{\min}} \quad (1.41)$$

where $I_{\max}$ and $I_{\min}$ are the maximum and minimum values of the intensity of the fringes; V has a maximum value of unity when $I_{\min}$ is zero.

We see that when $\tau = 0$, the interference pattern represented by equation (1.40) has unity visibility. Since $M(t)$ is periodic, with period $1/\Delta f$, the fringes have also unity visibility when

$$\tau = n/\Delta f, \qquad n = 1, 2, \ldots$$

When the value of τ departs from these values, the visibility generally decreases, and falls to a minimum value when

$$\tau = (n + \tfrac{1}{2})/\Delta f \qquad n = 0, 1, 2, \ldots$$

Thus the visibility of the fringes obtained with a laser beam varies periodically as the path difference between the two beams is altered.

The coherence length of a light source is generally defined as the distance in which the fringe visibility falls to $(1/e)^2$. In the case of a laser source, this is rather misleading, since the visibility may fall to this value in quite a short distance and increase to unity again when the path difference is equal to a cavity length.

Some lasers (known as single mode lasers) are made to produce only a single frequency in the sense defined above by introducing a tuning element known as an etalon. This causes a reduction in the output laser power since it eliminates the other modes. For this reason it is only used when continuous coherence is required. This wave has a coherence length of the order of tens of metres. (In practice, the precise value depends on the stability of the laser and can be increased if suitable precautions are taken.)

The output of a pulsed laser consists of a short pulse of light. The duration of the pulse may be from 10^{-6} to 10^{-12} s. The laser pulse can be described by equation (1.36), unless the laser is operated in a single mode configuration when it has a similar coherence length to the continuous laser source.

Interference can be observed using sodium, mercury and other conventional light sources. The coherence length of such sources is generally at best of the order of millimetres; a single colour filter is required to give even this much coherence.

1.5 Ray optics

In this section, some aspects of the behaviour of light energy which are most simply explained by using the ray model of light propagation are discussed. In particular the behaviour of a light beam when it meets a boundary between two materials with different light transmission properties is discussed. This leads to a description of the properties of mirrors, beamsplitters and lenses. In the last section, the use of such components to make practical devices for producing interference fringes (i.e. interferometers) is described.

1.5.1 *The ray model of light propagation and its application to reflection and refraction at boundaries*

In the ray model of light, the light energy is considered to travel in straight lines when travelling through a medium of constant optical transmitting properties. The direction of the ray at any point can be taken to be perpendicular to the direction of the wavefront (this is the case only when the material is optically isotropic – i.e. its properties are the same in all directions – only such materials are discussed in this book). When the light is incident on a boundary between two materials with different light transmitting properties, the ray is, in general, partly reflected and partly transmitted as shown in Figure 1.6. The incident ray R is incident at an angle i to the surface normal N. The reflected ray R' forms an equal but opposite angle i' to the normal N, and lies in the plane containing R and N. The transmitted ray lies in the same plane, the angle between T and N being r where r is given by the relationship known as Snell's Law:

$$n_1 \sin i = n_2 \sin r \tag{1.42}$$

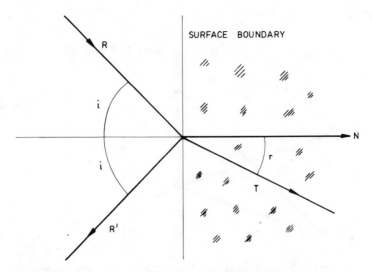

Fig. 1.6 Reflection and refraction at a plane boundary

where n_1 and n_2 are constants known as the refractive indices for the materials on each side of the boundary. It can be shown that n, the refractive index, is inversely proportional to the velocity of light in the material. The refractive index of a vacuum is taken to be unity and since the velocity of light in air is very close to that in a vacuum, the refractive index of air is often also taken to be unity. The velocity of light in any medium is always slower than the velocity of light in a vacuum, so that the refractive index is always greater than or equal to unity.

We see that when $n_2 < n_1$, the value of sin r given by equation (1.42) can be greater than unity. For some values of i, there is no transmitted ray; all the light is reflected at the first boundary. The minimum angle at which this occurs is known as the critical angle, and is given by

$$i_c = \sin^{-1} \frac{n_2}{n_1}$$

It can be shown that the frequency of the light as it travels through any medium must remain constant. Thus, the wavelength λ of light in a medium of refractive index n is related to its wavelength λ_0 in air by

$$\frac{\lambda}{\lambda_0} = n \tag{1.43}$$

With this change of wavelength is associated a change of phase. Thus two beams which have got the same initial phase and travel through

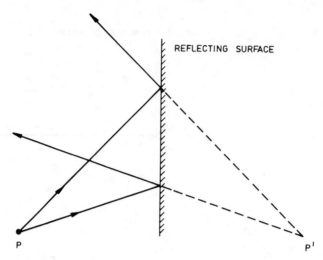

Fig. 1.7 A spherical wavefront reflected at a plane boundary

equal lengths of materials of different refractive indices will emerge with different phases.

Many materials impose considerable attenuation of the amplitude of the transmitted rays. However, glass, a few other solids and many liquids transmit light with very little attenuation over the distances of interest here.

1.5.2 *Reflection at a plane boundary*

A wavefront can be considered to be composed of a set of rays. When such a set of rays is incident on a plane boundary, each ray is reflected into an equal and opposite angle to its angle of incidence. The set of rays which is reflected from the boundary will appear to come from behind the boundary so that the shape of the original wavefront is maintained. For example, if a spherical wavefront with origin at P (Figure 1.7) is incident on the boundary, the reflected wave will appear to come from P' which is located on the normal from P to the boundary at an equal distance on the opposite side of the boundary.

This explains the action of mirrors: the light from an object appears as if it comes from an equivalent object located behind the boundary.

When a ray travels through a slab of material of refractive index n whose opposite faces are parallel to one another (see Figure 1.8), it can be shown from equation (1.42) that the light which emerges from the other side of the medium is travelling with the same orientation, but is

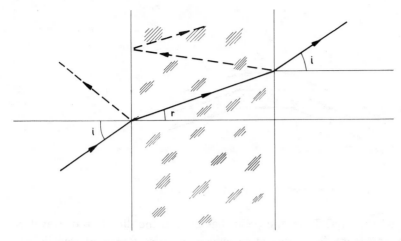

Fig. 1.8 Transmission of a ray through a slab of material that has parallel faces

displaced with respect to the incident ray. Thus, both the reflected and transmitted parts of a wavefront incident on such an element maintain the shape of the original wavefront. Such a slab can therefore be used to divide an incident wavefront into two wavefronts similar in shape to the original wavefront, one of which is reflected and the other transmitted. Such a device is known as a beamsplitter. The ratio of reflected to transmitted light intensity can be altered by coating one face with a semi-reflecting material. If one face is coated with a layer of metal which is several wavelengths thick, the surface will reflect all the incident light, thereby constituting a mirror.

It should be noted that in the element shown in Figure 1.8, another reflection (shown dashed) will occur at the right hand boundary. The ray reflected here will subsequently be partially reflected on the left hand side, and so on.

1.5.3 *Refraction of light by a prism, and by a spherical surface*

Consider a slab of glass in which one face is tilted at an angle with respect to the other as shown in Figure 1.9. A plane wavefront, which may be considered as a set of parallel rays, is incident on one face. Within the material, the rays are refracted towards the normal direction. The rays will still be parallel to one another, they will be bent again on emerging from the prism and will again be mutually parallel. It can be shown using equation (1.42) that the phase change is constant

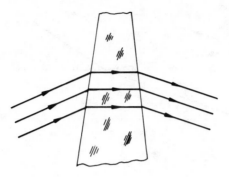

Fig. 1.9 Transmission of a ray through a wedge

for all the rays. Thus, the prism has altered the direction of travel of the wavefront. The amount by which the direction is altered clearly depends on the angle between the two faces of the prism; the smaller the angle, the less the alteration in direction.

Consider now a piece of glass of which one face is a section of a spherical surface and the other is flat – see Figure 1.10. The spherical surface can be considered to be a prism of varying angle. A ray which is incident in a direction which is perpendicular to the plane face, and passes through the centre of curvature of the spherical surface emerges from the element in the same direction. Any other ray parallel to this direction will be deflected towards it by the element, the amount of the deflection being increased as the ray moves further from the centre. The shape of the emergent wavefront is clearly altered, and the change may be calculated by following particular rays through the system. It can be shown that all such rays converge to a single point when their separation from the central ray is very much less than the radius of curvature of the spherical surface (the small angle or paraxial approximation). Thus,

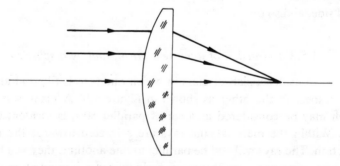

Fig. 1.10 Transmission of rays through an element having one spherical surface and one flat surface

a plane wavefront is converted into a converging spherical wavefront. This optical component is known as a lens.

The point of convergence is known as the focal point of the lens and the focal length is the distance between this point and the centre of the lens. The line which joins the centre of the lens to its centre of curvature is known as the principal axis of the lens and the plane perpendicular to the principal axis in which the focal point is located is called the focal plane. (Another quantity frequently used in the discussion of lenses is the F-*number* or the *numerical aperture* – this is the ratio of the focal length of the lens to its diameter.)

The focal length of a lens is given by

$$1/f = (n-1)/R \qquad (1.44)$$

where n is the refractive index of the material and R is the radius of curvature of the spherical surface.

It can be shown that all the rays will be in phase when they arrive at the focal point, since each ray has travelled through a different thickness of glass.

A lens may have two spherical surfaces; in this case, the focal length is

$$1/f = (n-1)\left(\frac{1}{R_1} - \frac{1}{R_2}\right) \qquad (1.45)$$

where R_1, R_2 are the radii of curvature of the two surfaces.

Equation (1.45) can be applied to a lens having a concave surface(s) by assigning a negative value to R_1 and/or R_2. A negative focal length implies that an incident plane wave appears to diverge from a point at a distance f behind the centre of the lens – see Figure 1.11. In this case, a plane wavefront is converted to a diverging spherical wavefront.

It can also be shown, again within the small angle approximation, that a plane wave which is incident at an angle θ to the principal axis of the lens – see Figure 1.12 – is focussed to a point P at a distance f from the lens. The height of P above the principal axis is

$$h = f\theta \qquad (1.46)$$

The path of any ray R incident on the lens can be found as follows.

A ray strikes the lens at a point X which is at a height x above the lens centre. A plane wavefront parallel to R is focussed to the point P in the focal plane at a height h, given by equation (1.46).

The transmitted ray R' travels in the direction XP. The angle ψ between R' and the principal axis of the lens can be calculated or may

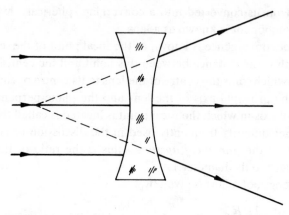

Fig. 1.11 Transmission of a plane wavefront through a lens which has concave faces

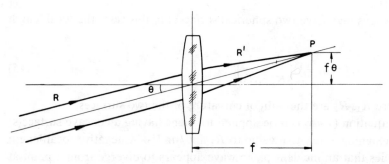

Fig. 1.12 Rays incident at an angle θ brought to a focus at the point P

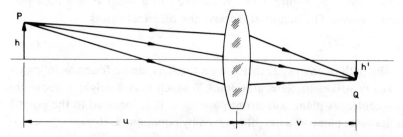

Fig. 1.13 All rays from the point P converge to the point Q where u and v satisfy the lens equation, (1.48)

be found graphically, and is given by

$$\psi = \theta - \frac{x}{f} \qquad (1.47)$$

Any ray incident on the lens from a point P (see Figure 1.13) can be shown using equation (1.35) or by a graphical construction, to intersect the line PQ at a fixed point Q, whose normal distance from the lens, v, is given by the equation

$$\frac{1}{u} + \frac{1}{v} = \frac{1}{f} \qquad (1.48)$$

where u is the normal distance from P to the lens.

It can also be shown that the heights h and h' of P and Q above the principal axes are given by

$$\frac{h'}{h} = \frac{-v}{u} \qquad (1.49)$$

Equation (1.48) is known as the lens equation. When a point source is on the left hand side of a convex lens at a distance $u > f$, it is brought to a focus on the right hand side. Each ray from the point source travels through the same optical path; all the rays arriving at the point of focus are in phase. An illuminated object on the left hand side of the lens can be considered as a set of point sources; since each point will be focussed by the lens, an image of the object is formed. From equation (1.49) we see that the image is inverted and its size is altered.

The image in this case is a real image, which would produce a recorded image on a photographic plate located at the position of the image. If $u < f$, a negative value of v is obtained. By tracing the rays, it will be seen that the light diverges from the lens in such a way that a magnified image of the object appears to exist behind the object. (This is the principle of a magnifying glass). Such an image is known as a virtual image. A negative value of v in the lens equation always implies a virtual image.

When a lens of negative focal length i.e. a concave lens, is used, the image formed is always virtual, and is always reduced in size.

The lens equation can be used to find the effect of a combination of lenses, by determining the position of the image formed by each lens in turn and using this image as an object for the next lens. If an intermediate image is located beyond the next lens in the system, a negative value should be assigned to u. When the small angle approximation is not valid the rays no longer converge to a point and a distorted image is obtained; this is known as aberration. To obtain an aberration-free image for larger angles of incidence, more complex lenses must be used. These generally consist of a combination of spherical elements of various radii of curvature and refractive indices. Such a lens can be designed using ray tracing; the distortion in the image produced by a given combination of lenses is found and the various parameters of the component lenses are adjusted until minimum image distortion is obtained.

In using the ray model of light propagation we have found that a point source of light is brought to a point focus in the image plane by an aberration-free system; it will be seen in Section 1.4 that due to diffraction effects, a point focus is never obtained. When, however, a lens or lens system is free of geometrical aberrations it is said to be 'diffraction limited'.

Since refractive index is a function of wavelength, the focal length of a lens varies with wavelength; this is known as chromatic aberration, and can also be compensated for by using a combination of lenses; for example, the achromatic doublet.

1.5.4 *Expansion and collimation of laser beams*

The output of a laser is in the form of an approximately plane wavefront whose diameter is of the order of millimetres. To illuminate an object larger than this, the beam must be expanded. This can be done using a lens – see Figure 1.14. The beam is brought to a focus at the focal point of the lens and subsequently expands. The rate at which it expands obviously depends on the focal length of the lens. The shorter the focal length the greater the angle subtended by the expanding beam at the focus, and hence the greater the expansion of the beam at a given distance. A microscope objective lens which has a very short focal length is often used for this purpose.

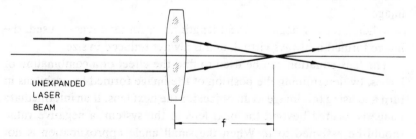

UNEXPANDED
LASER
BEAM

f

Fig. 1.14 Expansion of a laser beam using a lens

A plane wavefront covering an area considerably larger than the diameter of the laser beam is frequently required in speckle and holographic interferometry. This can be made by placing a second lens of focal length f_2 so that its focus coincides with that of the first lens of focal length f_1. The output beam is then a plane wavefront; the ratio of the diameters of input to output beam is given by f_2/f_1. (See, for example, Figure 2.11.)

1.5.5 *Interferometers*

Interferometry enables measurements to be made to an accuracy of the order of one wavelength of the light used and in some cases to a tenth or even a hundredth of a wavelength. It is thus a very important measuring tool. In this section some interferometers and their applications are discussed.

Interference fringes are obtained by combining two wavefronts. If the wavefronts are plane, the form of the fringes is simple. (Section 1.3.4) Equation (1.26) shows that unless the angle θ between the two beams is small, the fringes will be very close together. To obtain fringes whose spacing is of the order of 0.1 mm, using a HeNe laser, θ must be $\sim 6 \times 10^{-3}$ radians. Many methods of combining two wavefronts at such small angles have been devised. One important such instrument is the Michelson interferometer which is shown in Figure 1.15.

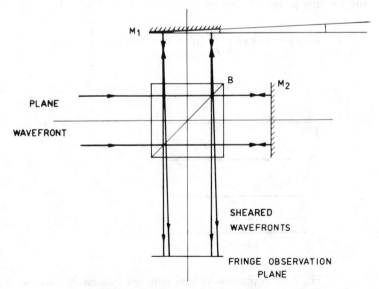

Fig. 1.15 A Michelson interferometer

An expanded and collimated laser beam is incident on a beam splitter B. One beam is reflected by M_1 and is partly transmitted by B to the viewing plane. The second beam is reflected by M_2 and partly reflected by B so that it combines with the first beam to give interference fringes. By varying the orientation of one of the mirrors M_1, M_2 the angle between the two beams can be altered, so that the spacing and orientation of the fringes can be altered. If θ is zero, the fringe spacing becomes infinite and what is known as a 'fluffed out fringe' is seen, where the intensity is uniform.

If one of the mirrors is not perfectly flat, it will distort the wavefront incident on it and thereby distort the fringe pattern. Observation of this distortion enables the departure from flatness of the mirror to be measured. This can be shown as follows.

In Figure 1.16, the mirror is in the xy-plane, and the plane wavefront is incident in the z-direction. Those parts of the surface which are not flat will reflect the incident wavefront so that its direction of propagation is tilted away from the z-direction. However, this angle will generally be very small for a nominally flat mirror, so that a ray can be assumed to be reflected back along its own path. The distance travelled by a given part of the wavefront depends on which part of the surface it is reflected from, so that the phase varies across the reflected wavefront. If the departure of the mirror from flatness is given by $\zeta(x, y)$ and the distance from the mirror to the viewing plane is z_0, the amplitude at the viewing plane is given by

$$U_1(r) = A_0 \exp\left\{\frac{4\pi i}{\lambda}[(\zeta(x, y) + z_0)] + i\psi_1\right\} \qquad (1.50)$$

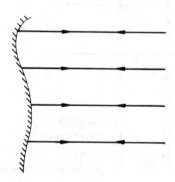

Fig. 1.16 A plane wavefront reflected by a non-flat mirror (departure from flatness exaggerated for clarity)

The amplitude of the second beam may be represented by

$$U_2(r) = A_0 \exp \left(\frac{4\pi i z_0}{\lambda} + i\psi_2 \right) \tag{1.51}$$

The resultant intensity is then given by

$$I = 2I_0 \left\{ 1 + \cos \left[\frac{4\pi}{\lambda} \zeta(x, y) + (\psi_1 - \psi_2) \right] \right\} \tag{1.52}$$

Fringes will be observed in the viewing plane, with maxima occurring when

$$\frac{4\pi}{\lambda} \zeta(x, y) + (\psi_1 - \psi_2) = 2n\pi \tag{1.53}$$

Thus the fringe pattern maps out the variation of the height of the mirror as contours of interval $\frac{1}{2}\lambda$; this demonstrates the ability of interferometers to map height variations with a sensitivity of one wavelength of light.

A modification of a Michelson interferometer known as a Twyman–Green interferometer is an instrument very commonly used for testing the accuracy of optical components such as lenses, mirrors and prisms. A description of this will be found in standard optics textbooks – see references (1)–(3).

A Michelson interferometer can be very easily set up using an unexpanded laser beam. In this case, an expanding lens is normally inserted in the viewing plane to increase the fringe spacing. Such a system provides a simple method of measuring the coherence properties of a laser, particularly when it has several longitudinal modes – Section 1.4.2. If one mirror is further away from the beamsplitter than the other, the two beams which are recombined in the viewing plane will have left the laser at different times. As the difference in path is increased, the two beams can no longer be considered to be a single wave of uniform frequency and the contrast of the fringes is reduced, and falls to a minimum when the path difference is equal to the laser cavity length. If the mirror is moved further back, fringes will once again be obtained (see Section 1.4.2). In any interference experiment (which includes all speckle and holographic experiments) the path difference between the two beams must be kept within the range where the fringe contrast remains good, or alternatively it must be equivalent to this distance plus an integral number of cavity lengths (see also Section 6.4.1).

Interferometry can also be used to measure changes in density which occur as the result of pressure or temperature changes of fluids, since

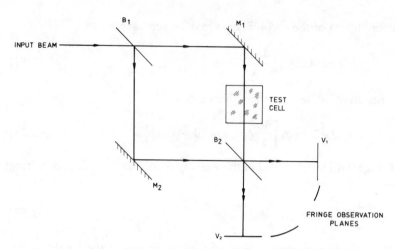

Fig. 1.17 A Mach–Zehnder interferometer

such changes produce changes in refractive index. A change in refractive index means that the wavelength of the light travelling through that medium is altered, so that the phase change of the light after travelling through the medium is altered. The Mach–Zehnder interferometer, shown in Figure 1.17 is frequently used for such measurements. The input beam is split by B_1. One beam travels via M_1 and B_2 to the viewing plane which in this case may be either at V_1 or V_2. The second beam travels via M_2 and B_2 to V_1 or V_2. The relative orientation of the beams can be adjusted using the various mirrors and beamsplitters. If a cell is included in one path containing a fluid, then the wavefront travelling through the cell can be described by

$$U(r) = A_0 \exp \left\{ \frac{2\pi i}{\lambda} \left[z_0 + \int_0^L n(r) \, dz \right] + i\psi_1 \right\} \tag{1.54}$$

where z_0 is the distance travelled outside the cell, L is the length of the cell, and $n(r)$ is the refractive index at the point (r) within the cell. The refractive index appears here since the wavelength of the light in the medium and that in air are related by equation (1.43). The phase term ψ_1 includes the phase change introduced by the cell windows and the beamsplitters which is assumed to be constant across the fringe field.

If the density of the fluid varies within the cell, the value of $\int_0^L n(r) \, dz$ will vary for different parts of the beam. Putting

$$\int_0^L n(r) \, dz = N(x, y)$$

the complex amplitude becomes

$$U(r) = A_0 \exp\left\{\frac{2\pi i}{\lambda}[z_0 + N(x, y)] + i\psi_1\right\}$$

The intensity of the two beams when combined is then given by

$$I = 2I_0\left\{1 + \cos\left[\frac{2\pi}{\lambda}N(x, y) + (\psi_1 - \psi_2)\right]\right\} \tag{1.55}$$

Thus, variations in $N(x, y)$, which can be seen as the integrated optical density along the cell, are shown as fringes occurring at intervals of

$$\Delta N(x, y) = \lambda \tag{1.56}$$

The advantage of using a Mach–Zehnder interferometer for this type of measurement is that the light passes through the cell only once so that the fringe pattern is more easily analysed.

1.6 Diffraction

When a light beam is incident on an aperture, an edge or a surface which is rough compared with the wavelength of the light, it behaves in a way which is not consistent with the ray model of light propagation. It is found under these circumstances that the light is spread out in a way not consistent with reflection and Snell's law of refraction. For example, a plane wavefront incident on an aperture – see Figure 1.18 – becomes divergent after passing through the aperture. This effect, which is very easily observed using laser light (e.g. by placing a hair in the beam), is known as diffraction.

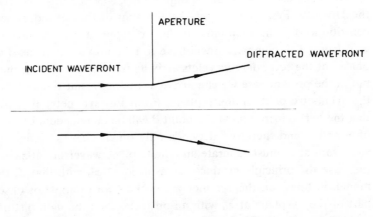

Fig. 1.18 Diffraction of a plane wavefront incident on an aperture

1.6.1 *The Huygens–Fresnel principle*

The Huygens–Fresnel principle was derived empirically to describe diffraction and was later confirmed analytically with the Fresnel–Kirchhoff diffraction formula and more generally by that of Rayleigh–Sommerfeld. This principle is as follows.

Each point on a wavefront acts as a point source of so-called secondary waves. The amplitude of the wavefront as it propagates is given by the sum of the amplitudes of these secondary waves. The amplitude of the secondary wavelets is a maximum in the direction of propagation and falls off to zero in the orthogonal direction. The amplitude contributed by the secondary source at a point P to a given point Q is described by an obliquity function $f(\eta)$ where η is the angle between the direction of propagation at P and the line PQ.

Different values of $f(\eta)$ are obtained in the Fresnel–Kirchhoff and the Rayleigh–Sommerfeld formulae. It will be assumed throughout this text, unless it is explicitly stated to the contrary, that the range of values of η in any given situation will be sufficiently small to treat $f(\eta)$ as a constant – i.e. that the angles of diffraction are small; this is known as Fraunhoffer or far field diffraction.

1.6.2 *Diffraction by a single aperture*

The diffraction principle can be used to describe the propagation of any wavefront. Consider first the case of a uniform, infinite plane wave propagating in the z-direction. Let the wavefront be at z_0 at time t_0. The light disturbance at a point P $(x, y, z_0 + \Delta z)$ will, according to the Huygens–Fresnel principle, be the sum of the secondary wavelet contributions from all points in the xy-plane at z_0 occupied by the wavefront. This summation will have to take into account the differing phases of the secondary wavelets arriving from each point on the front but in the present case we can arrive at the net result simply by noting the symmetry of the infinite plane. From this symmetry it is obvious that the net contribution at the point P will be independent of the value of x and y and therefore depends only on Δz, i.e. the infinite plane wavefront at z_0 must generate an infinite plane wavefront at $z_0 + \Delta z$. In this case the principle produces a result identical with that of the ray model. If, however, this symmetry is broken, for example by obscuring part of the xy-plane at z_0 with an obstacle, then the light disturbance at $(x, y, z_0 + \Delta z)$ will in general be a continuous function of x and y

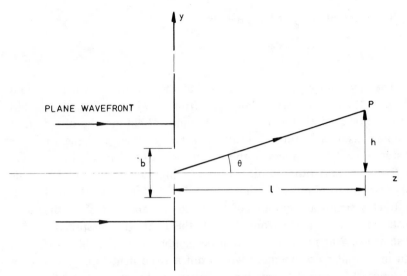

Fig. 1.19 Diffraction of a plane wavefront incident on an infinitely long slit of width b

which depends on the relationship of the missing secondary wavelets to the net sum at each point and the wavefront is no longer plane.

Consider now the case of a plane wavefront travelling in the z-direction incident on an infinitely long slit of width b which lies along the x-axis, with its centre at the origin – see Figure 1.19. A point P of coordinates $(0, h, l)$ receives light from all points in the aperture. The angle θ between the centre of the slit and P is given by $\tan \theta = h/l$. The contribution $\mathrm{d}U$ to the amplitude at P due to the waves from a line element $\mathrm{d}y$ centred on y can be written as

$$\mathrm{d}U = u_0 \exp\left[\frac{2\pi\mathrm{i}}{\lambda}(l_0 - y \sin\theta)\right]\mathrm{d}y \tag{1.57}$$

where l_0 is the distance from the centre of the aperture to P, and it is assumed that θ is small.

The total complex amplitude at P is then given by

$$U(\mathrm{P}) = u_0 \int_{-\frac{1}{2}b}^{+\frac{1}{2}b} \exp\left[\frac{2\pi\mathrm{i}}{\lambda}(l_0 - y \sin\theta)\right]\mathrm{d}y$$

$$= u_0 b \exp\frac{2\pi\mathrm{i}l_0}{\lambda} \operatorname{sinc}\left(\frac{\pi b \sin\theta}{\lambda}\right) \tag{1.58}$$

where

$$\operatorname{sinc} x \equiv \frac{\sin x}{x}.$$

The total intensity at P is given by

$$I = I_0 \, \text{sinc}^2 \left(\frac{\pi b \sin \theta}{\lambda} \right) \tag{1.59}$$

The ray theory of light predicts that the wavefront will propagate through the aperture as a plane wavefront of width b. Equation (1.59) shows, however, that a significant amount of light may be observed at points considerably further away from the axis than $\frac{1}{2}b$. Thus, the light disturbance is not confined to the geometrical area of the slit, but appears to be bent into other directions according to equation (1.59). This phenomenon is known as diffraction. For instance, a 10 μm slit will diffract a significant amount of light into an angle of 6° so that at a distance of one metre from the slit, the light will be spread over a distance of 200 mm in a direction perpendicular to the slit. A 100 μm slit in a similar arrangement would diffract the light to extend over a distance of 20 mm. The smaller the aperture, the greater the diffraction effect.

If the slit is replaced by a circular aperture the complex amplitude at an equivalent point will be given by an expression similar to equation (1.58) except that the integration is performed over the two variables r and θ describing the circular area.

The resultant complex amplitude at a point P at an angle θ from the axis and at a distance l from the aperture of diameter b is given by

$$U(\text{P}) = A_0 \exp \left(\frac{2\pi i l \cos \theta}{\lambda} \right) \left| \frac{J_1(\pi b \sin \theta / \lambda)}{(\pi b \sin \theta / \lambda)} \right| \tag{1.60}$$

where J_1 is a first order Bessel function – see Section 8.5 of reference (1).

$J_1(x)/x$ has a maximum value of 0.5 at $x = 0$, and an infinite number of zeros and maxima whose values fall off rapidly as $|z|$ increases. The intensity is given by

$$I = I_0 \left[\frac{2 J_1(\pi b \sin \theta / \lambda)}{(\pi b \sin \theta / \lambda)} \right]^2 \tag{1.61}$$

where the 2 is included in the bracket so that this function normalizes to unity at zero.

Thus, the intensity has a maximum value when $x = 0$. $J_1(x)$ equals zero when $x = 1.22\pi$, giving zero intensity when

$$\sin \theta = \frac{1.22 \lambda}{b} \tag{1.62}$$

We have so far considered only diffraction by an aperture in an opaque screen. It is very important to note, however, that any alteration of either the phase or the amplitude of the propagating wavefront will give rise to diffraction. If the screen discussed in the previous section is removed, and an opaque circular disc put in place of the circular aperture, the amplitude at the point P is now given by

$$U(P) = U_0 \exp i\, kz - U_{\text{diff}}(P)$$

where U_0 is the amplitude of the undiffracted wave, and $U_{\text{diff}}(P)$ is the amplitude of the wave which is diffracted by the circular aperture and is given by equation (1.60). We see that the transmitted amplitude is modified by the aperture. When θ is large the diffracted amplitude $U_{\text{diff}}(P)$ becomes very small, but when θ is of the order specified by equation (1.62), the amplitude is significantly modified. We see that the angles at which the minima and maxima occur in the case of an opaque disc correspond to the angles of the maxima and minima in the case of the aperture. Thus the structure of the diffracted light is similar in the two cases.

Consider now the introduction of a circular disc which alters the phase of the wave by a factor δ. In this case the amplitude at P is given by

$$U(P) = [U_0 \exp i\, kz - U_{\text{diff}}(P)] + U_{\text{diff}}(P)\exp(i\delta)$$

where the first two terms represent the amplitude of the light when the circular area is totally obstructed, and the last term represents the light diffracted by the aperture to which the additional phase term $\exp i\delta$ has been added. We can write

$$U(P) = U_0 \exp ikz + U_{\text{diff}}(P)\,[(\exp i\delta) - 1]$$

We see that the transmitted amplitude is modified by the diffraction term $U_{\text{diff}}(P)$ and again the structure has a form similar to that of the circular aperture. An equivalent result will be obtained if the amplitude is modified over the circular area.

This is a very important result in the discussion of holography. When a wave is modified over a given area, the structure of the resulting diffraction pattern is similar whether the modification consists of a variation in phase, or amplitude of the propagating wave or both.

The diffraction described by equation (1.61) is known as the Airy diffraction pattern, and the central bright disc is known as the Airy disc. Using this, we can explain why a lens does not give a perfect point image for an incident plane wave, as follows:
A plane wave is incident on the lens L of diameter D. The lens or the lens aperture (an aperture is frequently placed in front of an imaging

forming lens to alter its imaging properties; a detailed discussion is outside the scope of this book but will be found in standard optics textbooks) diffracts the incident wave so that it is no longer plane, but contains components travelling at angles θ whose amplitudes are described by equation (1.60). A wave at angle θ will be brought to a focus at a point given by equation (1.46), and so the intensity of the light in the focal plane will be described by equation (1.61). Instead of a point, a bright disc surrounded by rings will be obtained. The larger the diameter of the lens or the lens aperture, the smaller the disc.

If two points in the object plane are very close together, their images will overlap; the two images will be indistinguishable as such if the points are sufficiently close together. Thus, the ability of a lens to resolve fine detail is limited by diffraction.

1.6.3 *Diffraction by two apertures*

The effect of two small slits located close together on an incident plane wavefront is interesting from a historic point of view since it was used by Young in 1801 in one of the first demonstrations of the wave nature of light. It is also of importance in the context of this book since it is used in speckle interferometry and speckle photography – Sections 3.2 and 3.6.3.

A plane wave is incident normally on a screen containing two slits of width b separated by a distance d, Figure 1.20. The complex amplitude of the light at a point P which is at a distance l from the centre of the screen and at an angle θ from the normal to the screen is given by an

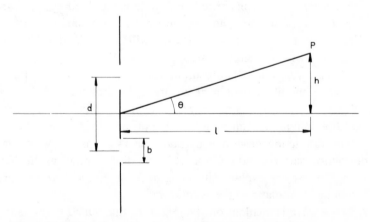

Fig. 1.20 Diffraction of a plane wavefront by two slits

expression similar to equation (1.58) except that the limits are altered:

$$U(P) = A_0 \int_{-\frac{1}{2}d-\frac{1}{2}b}^{-\frac{1}{2}d+\frac{1}{2}b} \exp \frac{2\pi i}{\lambda} (l_0 - y \sin \theta)\, dy$$

$$+ A_0 \int_{\frac{1}{2}d-\frac{1}{2}b}^{\frac{1}{2}d+\frac{1}{2}b} \exp \frac{2\pi i}{\lambda} (l_0 - y \sin \theta)\, dy$$

Evaluation of this equation gives

$$U(P) = 2A_0 b \frac{\sin (\pi b \sin \theta / \lambda)}{(\pi b \sin \theta / \lambda)} \exp (2\pi i l_0 / \lambda) \cos (\pi d \sin \theta / \lambda)$$

(1.63)

The intensity may be written as:

$$I = I_0 \operatorname{sinc}^2 \left(\frac{\pi b \sin \theta}{\lambda} \right) \cos^2 \left(\frac{\pi d \sin \theta}{\lambda} \right)$$

(1.64)

The variation in this intensity as a function of θ is shown in Figure 1.21, and is seen to be similar to that which would be obtained if a single slit of width b were used, but modified by the cosine term which represents the interference between the waves originating from the two slits. The relative spacing of the two patterns depends on the relative values of b and d.

Two circular apertures will give a diffraction pattern which consists of the Airy diffraction pattern (equation 1.61) modified by an interference term.

1.6.4 *The diffraction grating*

A diffraction grating is a device which modifies the transmission of a wavefront in a periodic manner, where many cycles of the

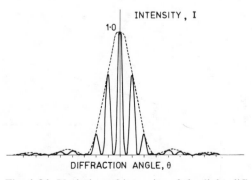

Fig. 1.21 Variation of intensity of the light diffracted by two slits, plotted as a function of angle of diffraction

transmission function occur over the complete wavefront. A simple example is a screen with a series of narrow equidistant slits. The effect of such a grating is to divide the incident beam into a set of beams travelling in discrete directions which are determined by the grating spacing and the wavelength of the light. This can be explained as follows.

Consider a grating made up of a set of slits of width b separated by a distance d, illuminated by a plane wavefront which is incident in the normal direction. Each slit produces a diffracted wavefront and each of these wavefronts interferes with all the others. The amplitude of the light at a point P (Figure 1.22) diffracted from the pth slit where $p = 0$ at the origin, is given by

$$U_p(P) = u_0 b \exp (2\pi i \, l_p/\lambda) \operatorname{sinc} [(\pi b \sin \theta)/\lambda] \qquad (1.65)$$

where l_p is the distance from the centre of the pth slit to the point P. It may be written as

$$l_p = l_0 + pd \sin \theta.$$

where l_0 is the distance from the origin to P.

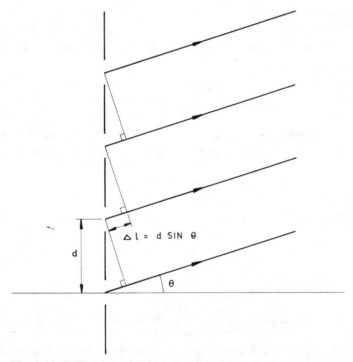

Fig. 1.22 Diffraction of light by a periodic grating

The total amplitude at P is then given by

$$U(\text{P}) = u_0 b \exp(2\pi i\, l_0/\lambda)\, \text{sinc}(q) \sum_{p=-M}^{+M} \exp\left[(2\pi i\, pd \sin \theta)/\lambda\right]$$

(1.66)

where $q = (\pi b \sin \theta)/\lambda$, and $N = (2M + 1)$ is the total number of slits. Putting

$$u_0 b \exp(2\pi i\, l_0/\lambda)\, \text{sinc}(q) = U_1,$$

we have

$$U(\text{P}) = U_1 \sum \exp\left[(2\pi i\, pd \sin \theta)/\lambda\right]$$

(1.67)

The summation is simply a geometric series whose sum can be written as

$$U(\text{P}) = U_1 \left\{ \frac{1 - \exp\left[2\pi i(2M + 1)d(\sin \theta)/\lambda\right]}{1 - \exp\left[(\pi i d \sin \theta)/\lambda\right]} \right\}$$

The intensity is proportional to $U(\text{P})U^*(\text{P})$ and can be shown to be given by

$$I = I_0 \frac{\sin^2\left[(\pi N d \sin \theta)/\lambda\right]}{\sin^2\left[(\pi d \sin \theta)/\lambda\right]}$$

(1.68)

This intensity is plotted as a function of θ in Figure 1.23. It has maximum values when

$$\frac{\pi N d \sin \theta}{\lambda} = (n + \tfrac{1}{2})\pi$$

and minima when

$$\frac{\pi N d \sin \theta}{\lambda} = n\pi,$$

except when $\dfrac{\pi d \sin \theta}{\lambda} = n\pi$

(1.69)

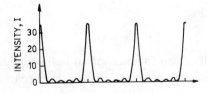

ANGLE OF DIFFRACTION, θ

Fig. 1.23 The intensity of the light diffracted by a periodic grating plotted as a function of angle of diffraction

In this case both sine terms are equal to zero and it is seen that since

$$\lim_{x\to 0} \frac{\sin Nx}{x} = N$$

the value of the intensity is a maximum at such points. The value of this maximum is considerably greater than the other maxima. Maxima satisfying equation (1.69) are known as principal maxima, and it is seen that they occur when

$$\sin \theta = \frac{n\lambda}{d} \tag{1.70}$$

Intense diffracted beams are observed at the angles defined by (1.69), while at other angles very little light is observed. The predominance of diffraction into the directions given by (1.69) can be understood by considering the light diffracted into the angle θ from equivalent points in two adjacent slits – see Figure 1.22. The path difference between these two components is given by

$$\Delta l = d \sin \theta$$

In general, each slit produces a component which is out of phase with the next one. Only when Δl equals an integral number of wavelengths will all the components be in phase: this corresponds to the condition given by equation (1.69). When $n = 0$, the beam is the zero order diffracted beam; when $n = 1$, it is the first diffracted order, and so on. The zero order beam is, of course, a beam travelling in the direction of the original wavefront.

As the value of N increases, it can be shown that the principal maxima increase and the secondary maxima decrease in height, and the widths of all the peaks decrease. The number of slits N is a measure of the total width of the grating and it is seen that the width of the individual peaks can be considered to be due to diffraction by the whole grating. The amplitude term U_1 contains the term

$$\text{sinc}\,(q) = \text{sinc}\left(\frac{\pi b \sin \theta}{\lambda}\right)$$

which corresponds to the diffraction function of an individual grating. The intensities of the higher order peaks decrease in accordance with this function.

It was shown in Section 1.6.2 that any modification of the propagating wave over a section of its wavefront gives rise to diffraction effects. Thus, if a periodic modification is introduced, the transmitted amplitude is found by replacing the sinc (q) term in equation (1.66) with the function which describes the diffraction of an individual element. The main diffracted beams will be at the same angles as defined by equation (1.70) but the distribution of the light will be altered.

It may be shown as follows that the spatial frequency distribution of the light diffracted by the grating is simply the Fourier transform of the function which describes the grating.

The amplitude of the light transmitted by the grating is given by

$$U_T = U_0 f(x) \exp \mathrm{i}(kz + \phi_0)$$

where $f(x)$ describes the transmission function of the grating and $U_0 \exp \mathrm{i}(kz + \phi_0)$ describes the normally incident wavefront. (For simplicity we consider a one-dimensional grating, but the argument is equally valid for a two-dimensional grating.) If we expand $f(x)$ as a sum of spatial frequency components, we have

$$U_T = U_0 \exp \mathrm{i}(kz + \phi_0) \int_{-\infty}^{+\infty} g(f_x) \exp (2\pi \mathrm{i} f_x x) \, \mathrm{d}f_x \qquad (1.71)$$

From Section 1.4.1, we see that the exponential terms in the integral represent plane waves travelling with direction cosines

$$n_x = \lambda f_x$$

and the diffracted waves represent the spatial frequency components of the grating. A result which is very important in holography can now be derived – namely the form of the diffracted light given by a grating whose transmission varies sinusoidally. The transmission $t(x)$ of the grating is written as

$$t(x) = t_0 \left[1 + \cos \left(\frac{2\pi}{d} x \right) \right]$$

where d is the spacing of the elements in the grating. The diffracted light is now described by

$$U_T = U_0 t_0 \exp \left[\mathrm{i}(kz + \phi_0) \right] [1 + \cos (2\pi x/d)]$$

which may be written as

$$U_T = \tfrac{1}{2} U_0 t_0 \exp \mathrm{i}(kz + \phi_0)[2 + \exp \mathrm{i}(2\pi x/d) + \exp -\mathrm{i}(2\pi x/d)]$$

This represents three plane waves, one travelling in the z-direction and the others at angles $\pm \theta$ given by

$$\sin \theta = \lambda/d$$

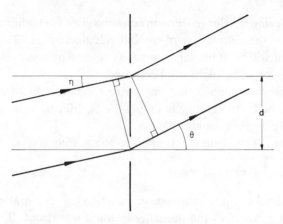

Fig. 1.24 Diffraction by a periodic grating of a plane wavefront
incident at an angle η to the grating

Thus, only the zero and the two first order diffracted waves appear
in this case.

So far in the discussion of diffracting gratings the incident wavefront
has been considered to be incident on the grating at normal incidence.
We will examine now the diffraction of a plane wavefront incident at
an angle η to the grating; see Figure 1.24. Consider the light diffracted
into an angle θ from equivalent points in two adjacent grating elements.
The path difference is given by

$$\Delta l = d[\sin \eta \mp \sin \theta]$$

Clearly the light from all the elements will be in phase when

$$d(\sin \eta \mp \sin \theta) = n\lambda \qquad (1.72)$$

Thus diffracted beams will appear at angles θ given by equation (1.72).
Again, this relationship applies to any grating whose grating spacing
is d.

1.7 Holography

Holography is a technique by which a wavefront can be recorded
and subsequently reconstructed in the absence of the original wavefront.
Observation of this reconstructed wavefront will give exactly the same
physical effect as the observation of the original wavefront, i.e. a three-
dimensional image is observed just as if the object were still present
and being illuminated in the same way as when the holographic recording
was made.

Although the principle of holography was proposed and demonstrated by Gabor (13, 14) in 1948 before the development of the laser, it has become a practical technique only with the availability of the coherent monochromatic light which is obtained from a laser and the almost simultaneous use of the offset reference beam principle by Leith and Upatnieks (15). This system forms the basis of most of the practical arrangements used for holography – see for example Figures 1.25, 2.11.

To make a good quality holographic recording, a laser source is required, but the reconstructed image may, in some circumstances, be observed with conventional light sources – either single colour sources, such as sodium or mercury lamps, or in some cases white light sources.

Holographic recording and reconstruction of a plane wavefront is discussed in 1.7.1, and of two plane wavefronts in 1.7.2. The extension to a complicated wavefront is given in 1.7.3.

1.7.1 *A two-beam interference grating as a hologram*

In Section 1.6.4. it was shown that a periodic grating illuminated by a wavefront gives rise to several diffracted wavefronts travelling in well defined directions which are defined by the grating spacing.

A grating of a given spacing may be produced by illuminating a photographic plate with two plane wavefronts incident at angles ψ_1 and ψ_2. The exposed and developed plate will have fringes whose spacing is given from equation (1.72) by

$$d = \frac{\lambda}{(\sin \psi_1 - \sin \psi_2)}$$

If the developed plate is illuminated by a plane wavefront travelling at an angle ψ_1, it is seen from equation (1.72) that one of the diffracted beams travels at the angle ψ_2. Thus the photographic grating reconstructs the wavefront which was incident at ψ_2 when illuminated by the wavefront incident at ψ_1. This photographic grating forms a simple hologram. In general, higher order diffracted waves will also appear at angles given by

$$\sin \theta = \frac{n\lambda}{d} - \sin \psi_1$$

It was shown in Section 1.6.4. that a sinusoidal grating gives only a zero and two first order diffracted waves; the photographic grating which is produced by the interference of two plane wavefronts can be made to approximate to this form by the use of suitable development and exposure times (see Section 6.5.2) so that the zero and first order beams predominate when it is used as a diffraction grating.

If the grating is illuminated by a beam travelling in the opposite direction to one of the original beams i.e. at an angle $(\pi - \psi_1)$, then a reconstructed beam is obtained travelling in the opposite direction to the other beam, i.e. at $(\pi - \psi_2)$. These beams are often referred to as being conjugate to the original beams.

It can be seen from equation (1.72) that if the plate is illuminated by a plane wavefront incident at an angle ψ_3, wavefronts should be reconstructed at an angle θ given by

$$\sin \theta = \frac{n\lambda}{d} - \sin \psi_3$$

Because the photographic emulsion has a finite thickness, however, it can be shown that, as ψ_3 departs from either ψ_1 or ψ_2, the amount of light diffracted decreases rapidly. This can be seen as follows.

The interference pattern in the holographic emulsion due to interference between the two plane wavefronts can be considered to be a set of planes parallel to the bisector of the wave directions normal to the plane in which these directions lie – see Section 1.3.4. A wave incident on the developed emulsion will undergo many partial reflections at each of the planes. When $\psi_3 = \psi_2$ or ψ_1, each of these reflections is in phase and the resultant amplitude is large; however, when ψ_3 departs significantly from one or other of these values, significant phase differences and hence, cancellations, occur between the individual reflected components, and the resultant diffracted intensity is very low.

In many instances in holography, the photographic plate is 'bleached' i.e. it is converted to a phase modulator by chemical processing which alters the refractive index of the material instead of its opacity. The plate gives the same form of diffraction as the transmission modulation plate (see Section 1.6.2), but does not attenuate the light and so gives more light intensity in the diffracted waves. (See Section 6.5.2 for a detailed discussion of holographic processing.)

1.7.2 A three-beam hologram

A photographic plate is exposed to three plane wavefronts incident at angles ψ_1, ψ_2 and ψ_3. The interference pattern obtained will consist of three superimposed sinusoidal gratings of spacing

$$d_{jk} = \left| \frac{\lambda}{(\sin \psi_j - \sin \psi_k)} \right|, \; j \neq k$$

It was shown in 1.6.4 – see equation (1.71) – that the light diffracted by a grating consists of the sum of a set of plane waves representing the spatial frequency components of the diffraction grating. Thus an incident wavefront is diffracted by the grating into a set of wavefronts which is the sum of the diffraction components of the individual gratings; seven diffracted wavefronts in all are obtained – a zero order beam travelling in the direction of the incident beam and two first order diffracted beams from each of the three gratings. When the beam is incident on the photograph at ψ_3, the d_{13} and d_{23} gratings give first order diffracted beams travelling in the directions of the original ψ_1 and ψ_2 beams; thus these two beams are reconstructed from the photographic grating. If these beams are to be observed, none of the other reconstructed beams should overlap them in space; this can be achieved by a suitable choice of ψ_1, ψ_2 and ψ_3. If the ψ_3 beam is offset from the ψ_1 and ψ_2 beams, the zero order beam and the two other first order beams diffracted from the d_{13} and d_{23} gratings will be separated in space from the reconstructed ψ_1 and ψ_2 beams. The angles θ at which the first order beams are diffracted by the d_{12} grating can be shown from equation (1.72) to be given by

$$\sin \theta = \pm |\sin \psi_1 - \sin \psi_2| - \sin \psi_3$$

so that whether or not the θ_{12} beams overlap the ψ_1, ψ_2 reconstruction depends on the values of ψ_1, ψ_2 and ψ_3. If, for example, $\psi_1 = 10°$, $\psi_2 = -10°$ and $\psi_3 = 45°$, one of the diffracted beams travels at an angle of $-21°$, and the other first order beam does not exist (the other value of $\sin \theta$ is greater than unity) so that the reconstructed ψ_1 and ψ_2 beams are well separated in space from the other beams.

If the emulsion is 'thick', the beams diffracted by the d_{12} grating from the incident ψ_3 beam will have very low intensity (see previous section) so that in this case it is necessary only to offset the ψ_3 beam from ψ_1 and ψ_2 reconstructions.

As with the two-beam grating, the use of a wavefront travelling in the opposite direction to that incident at ψ_1 reconstructs two beams travelling in the opposite directions to ψ_2 and ψ_3.

1.7.3 *Holography with a general wavefront*

Since any wavefront can be considered to be made up of a set of plane wavefronts – see 1.4.1 – it should now be clear how a holographic recording of such a wavefront is made and reconstructed. An object D is illuminated by an expanded laser beam, Figure 1.25. The wavefront scattered from the object D which is incident on the hologram

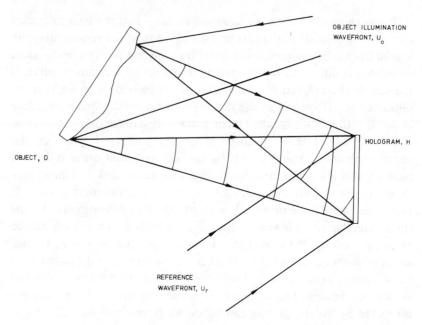

Fig. 1.25 An arrangement for making a holographic recording of the light scattered from an object

plate H is the sum of a set of plane waves which can be described by equation (1.29) as

$$U(x, y, 0) = \iint u_0(f_x, f_y) \exp 2\pi i(f_x x + f_y y) \, df_x df_y \qquad (1.73a)$$

where $\lambda f_{x,y} = n_{x,y}$, the direction cosines of the individual plane waves.

The plate is also illuminated by a plane wavefront which is known as the reference beam, and whose complex amplitude is given by

$$U_r(x, y, 0) = u_r \exp \frac{2\pi i}{\lambda}(\boldsymbol{n_r \cdot r}) \qquad (1.73b)$$

Each of the plane waves in equation $(1.73a)$ interferes with the plane wavefront described by equation $(1.73b)$ to give a grating of a particular spacing on the plate.

When the plate is developed and then illuminated by the reference wavefront (see Figure 1.26) it will reconstruct all the plane wavefronts described by equation (1.73). If the plate is viewed from the right hand side, a wavefront is seen which corresponds exactly to the wavefront which was scattered by the object D. If the reference beam is offset from the object beam, the zero order beam is separated from the

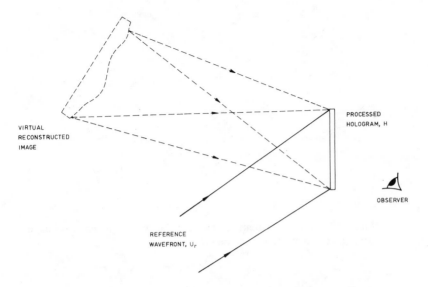

Fig. 1.26 An arrangement for viewing the reconstructed virtual
image obtained using the original reference beam

reconstructed object beam. The interference of the object beam with
itself will give rise to additional gratings which will diffract the incident
beam; these beams constitute noise in the reconstruction. Again suitable
choice of reference and object beam directions ensures that they do not
overlap the reconstructed beam; when the emulsion is thick, they will
in any case be of very low intensity.

Thus, when the developed hologram is illuminated by the original
reference beam and viewed from the other side, an image is seen which
is the same as that which would be seen by viewing the original object
from the same position with the same illumination; the image is of course
a virtual one.

In addition to this virtual image, a real image is obtained by the
diffraction of the reference beam into the $n = -1$ order. This image has
rather peculiar properties – it is known as a pseudoscopic image, which
can be roughly defined as being 'inside out'.

The reference beam used does not need to be a plane wavefront. If
a spherical wavefront is used, the intensity variation on the plate can
be considered to be made up of the gratings formed by the plane waves
from the object and the plane waves comprising the spherical wave.
Reillumination by the spherical wave reconstructs all the original object-
plane waves.

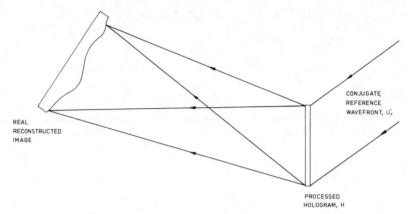

Fig. 1.27 An arrangement for viewing the reconstructed real image using a conjugate reference beam

If the hologram is illuminated by a wavefront travelling in the opposite direction to the original reference wavefront (see Figure 1.27), a true real image of the object is reconstructed at the position of the original object.

The holographic process may be represented mathematically as follows. Let the object and reference beam amplitudes at the hologram plate be given by U_0 and U_R respectively. The intensity may be written as

$$I \propto U_o U_o^* + U_r U_r^* + U_o U_r^* + U_o^* U_r$$

The transmission of the developed hologram plate can then be represented by

$$t = 1 - \alpha I$$
$$= 1 - \alpha[U_o U_o^* + U_r U_r^* + U_o U_r^* + U_o^* U_r] \qquad (1.74)$$

where α is a constant which is determined by the recording medium and processing, (see Section 6.3.1).

When the plate is illuminated by the beam U_r, the transmitted amplitude is given by

$$U_T = U_r - \alpha(U_o U_o^* U_r + U_r U_r^* U_r + U_o U_r^* U_r$$
$$+ U_o^* U_r U_r)$$

We see that the fourth term, $U_oU_r^*U_r$, may be written as

$$|U_r^2|U_o$$

when $U_r = u_r \exp i\, \phi_0$, so that this term represents the reconstruction of U_o.

The other terms represent the zero, first order and noise terms.

It has been assumed in this discussion that spatial frequency response of the holographic recording medium is sufficient to resolve the finest of the gratings given by the object–reference beam interference. It has also been assumed that the response of the recording medium is linear, i.e. that the modification of an incident wave by the developed plate is proportional to the intensity of the original interfering beams.

These conditions can be approximated very well using the appropriate holographic recording media and exposure and development techniques; these are discussed in Chapter 6. When they are not completely satisfied, the holographic image loses intensity and clarity, and optical noise components are introduced.

It has also been assumed that the reconstructing reference beam is exactly equivalent to the original reference beam. This is a requirement of holographic interferometry, but a holographic image is obtained even when it is not satisfied. However, the image may be distorted and magnified or demagnified as well as being reduced in intensity. The greater the departure from the original reference beam the greater these effects.

1.8 The speckle effect

Anyone who has seen a surface illuminated by a laser beam will have observed the curious granular appearance of such a surface. This is known as the speckle effect. A photographic image of a laser-illuminated surface is shown in Figure 1.28 and the speckle can be clearly seen. A similar effect occurs in coherent radar and ultrasonic imaging.

The speckle effect occurs only when the surface is optically rough, i.e. its height variation is of the order of, or greater than the wavelength of the illuminating light. When such a surface is illuminated by a laser beam, the intensity of the scattered light is found to vary randomly with position – this is known as objective speckle.

When a rough surface is illuminated by laser light and an image of the surface is formed, the image shows a similar random intensity variation, but in this case the speckle is called 'subjective'. The reason for this distinction should become clear in the next two sections.

Fig. 1.28 A subjective speckle pattern obtained by forming the image of an optically rough surface illuminated by a laser

1.8.1 *Objective speckle*

When a surface is illuminated, each point on the surface can be considered to absorb and re-emit the light, thus acting as a source of spherical waves similar to Huygens–Fresnel secondary waves. The complex amplitude of the scattered light at any point in space is given by the sum of the amplitudes of the contributions from each point on the surface. Consider a surface which is located in the xy-plane (see Figure 1.29). The surface height at a point (x, y) is given by $\xi(x, y)$. The complex amplitude at a point $P(r)$ of the light scattered from the surface is the sum of the components scattered from the whole surface. This may be written

$$U(r) = k \int\!\!\!\int\limits_{-\infty}^{+\infty} u(x, y) \exp\left[\frac{2\pi i}{\lambda} G\xi(x, y)\right] dx\, dy$$

where k is a constant, $u(x, y)$ represents the complex amplitude of the light incident at (x, y) and G is a geometric factor associated with the

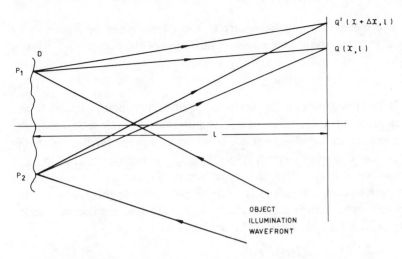

Fig. 1.29 The scattering of light by a rough surface

illumination and viewing directions which can be assumed to be constant when P is far away from the surface. Since the surface height is varying randomly by λ or more, the phase terms $G\xi(x, y)$ will also vary by amounts of λ or more. Hence the resultant amplitude at P is described as what is sometimes known as the 'drunkard's walk' – a set of vectors of random phase which when added together give a random resultant amplitude. The total amplitude has a value which varies between zero and a maximum value determined by the magnitude and phase of the individual amplitudes. As the point P is varied, the resultant amplitude and hence, intensity, will have a different random resultant value. It is this random intensity variation which is the speckle effect.

The statistical properties of speckle have been calculated by Goodman (16). The derivations are rather complex, and only the results will be quoted here.

The probability that the intensity I at a point, lies between I and $I + dI$ is given by

$$P(I)dI = \frac{1}{\langle I \rangle} \exp \frac{-I}{\langle I \rangle} \tag{1.75}$$

where $\langle I \rangle$ is the intensity of the speckle pattern, averaged over many points in the scattered field.

From (1.75) it can be shown that the probability $P(I')$ that the intensity is greater than some value I' is given by

$$P(I') = \exp \frac{-I'}{\langle I \rangle} \tag{1.76}$$

The mean value of the square of the intensity can be shown to be equal to $2\langle I\rangle^2$, so that the standard deviation of the intensity σ_I is given by

$$\sigma_I^2 = \langle I^2\rangle - \langle I\rangle^2 = \langle I\rangle^2 \qquad (1.77)$$

If the intensities of the scattered light at two points P and P' are compared, it is clear that when these points are very close together, the two intensities are closely related, but as they move further apart, the intensities become different. While these fluctuations do not have a fixed frequency, it is seen that the separation of the minima (or maxima) in the pattern falls within a fairly narrow range. While this mean speckle size cannot be quantified, it can be related to the autocorrelation function $R(r_1, r_2)$ of the intensity distribution, defined as

$$R(r_1, r_2) = \langle I(r_1)I(r_2)\rangle$$

where again the averaging is performed over many speckles. When $r_1 = r_2$, $R = \langle I^2\rangle$. As $(r_1 - r_2)$ increases, the intensities $I(r_1)$ and $I(r_2)$ are no longer the same, and eventually become totally unrelated to one another. In this case, we have:

$$R(r_1, r_2) = \langle I(r_1)\rangle\langle I(r_2)\rangle$$

which may of course be written

$$R(r_1, r_2) = \langle I(r)\rangle^2$$

The distance at which the intensities are unrelated provides a reasonable estimate of the 'size' of the speckles.

Goodman (16) has derived an expression for the autocorrelation function of the scattered intensity where the surface is illuminated by a uniform intensity beam of dimension such that an area of dimensions $L \times L$ is illuminated as

$$R(\Delta x, \Delta y) = \langle I\rangle^2\left[1 + \mathrm{sinc}^2\left(\frac{L\Delta x}{\lambda z}\right)\mathrm{sinc}^2\left(\frac{L\Delta y}{\lambda z}\right)\right] \qquad (1.78)$$

where z is the distance between the viewing and object planes, and $(\Delta x, \Delta y)$ are the x- and y-coordinates of the vector $(r_1 - r_2)$ representing the change in viewing position.

The average 'size' of a speckle can be taken as the value of Δx (or Δy) for which the sinc function first becomes zero, given by:

$$(\Delta x)_s = \frac{\lambda z}{L} \qquad (1.79)$$

The physical significance of this can be understood as follows.

The illuminated area is bounded on the x-axis by points P_1 and P_2 (see Figure 1.29). All points of the object contribute to the intensity at a given point $Q(x, l,)$ in the viewing plane. As the position of Q is varied, the amplitudes of the individual components vary very slowly, but their phases relative to one another alter rapidly.

The path difference from the points P_1 and P_2 to Q is given by:

$$s = (P_1Q - P_2Q) \simeq \frac{xL}{z} + \frac{1}{2}\frac{L^2}{z}$$

The difference in path from P_1 and P_2 to an adjacent point $Q'(x + \Delta x, l,)$ is given by

$$(P_1Q' - P_2Q') \simeq \frac{xL}{z} + \frac{1}{2}\frac{L^2}{z} + \frac{\Delta xL}{z}$$

Thus, the change in relative path to Q and Q' is given by:

$$\Delta s = \frac{\Delta xL}{z}$$

If Δl is considerably less than a wavelength, the relative phase of all the components will be approximately the same. If, however,

$$\frac{\Delta xL}{z} \simeq \lambda \tag{1.80}$$

the phases will be sufficiently different that the intensity at Q' will be unrelated to that at Q. Equation (1.80) corresponds to equation (1.79), the result obtained by Goodman using the autocorrelation function of the intensity distribution.

We see from (1.80) that the size of the speckles observed in the light scattered by a rough surface at a given distance from the surface increases as the area illuminated decreases.

The speckle pattern can be considered to be made up of a set of gratings of varying spatial frequencies; the maximum spatial frequency, f_{max}, is that of the grating formed by the interference of the light scattered from the edges of the illuminated area; the value of f_{max} is clearly related to the size of the illuminated area and the distance from the object to the viewing position.

This type of speckle is called objective because its scale depends only on the plane in space where it is viewed, and not on the imaging system used to view it. This is not the case for image-plane speckle which is discussed in the next section.

1.8.2 *Image-plane or subjective speckle*

When an image is formed of an object which is illuminated by laser light, the intensity of the image varies randomly – i.e. it is 'speckled'.

The parameters which described the probability distribution of the speckle intensity at a given point are the same as those of the objective speckle – equations (1.75)–(1.77).

The spatial distribution of the speckle is, however, determined by the diffraction limit of the imaging system. This is seen as follows.

A point P_1 in the object, Figure 1.30, forms a diffraction pattern centred on the point Q. The amplitude distribution is of the form described by equation (1.60) and the light from P_1 has a random phase associated with the random variation of the surface height. The point Q is also illuminated by points adjacent to P_1, since such points produce diffraction patterns which overlap onto Q. These diffraction patterns also possess random phases due to the surface height variation. A point P_2, which is located so that the first minimum of its diffraction patterns coincides with Q, makes no contribution to the complex amplitude of the light at Q.

A point further away from P_1 than P_2 will contribute a small amount of light to the amplitude at Q but since the secondary maxima of the diffraction pattern are very much smaller than the primary maxima, such contributions can be neglected. Thus, the intensity of the light at Q is made up of contributions from an area in the object centred around P_1 whose diameter d_{obj} is given by twice the distance between P_1 and P_2.

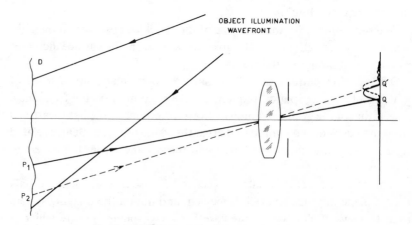

Fig. 1.30 Speckle pattern formation in the image plane of a lens

The distance QQ' can be found from equation (1.62) to be

$$QQ' = \frac{1.22\lambda v}{a}$$

where a is the diameter of the viewing lens aperture and v is the distance from the lens to the image plane. The 'size' of the speckles can be taken to be of the order of twice this quantity given by

$$d_{sp} \simeq \frac{2.4\lambda v}{a} \tag{1.81}$$

The distance $P_1 P_2$ which is the radius of the element of the object area which scatters light to the point Q is given by

$$(r_s)_{obj} = \frac{1.22\lambda u}{a} \tag{1.82}$$

where u is the object-to-lens distance.

Goodman (16) has derived an expression for the autocorrelation function of the image-plane speckle

$$R(r) = \langle I \rangle^2 \left[1 + 2J_1\left(\frac{\pi a r}{\lambda v}\right) \Big/ \left(\frac{\pi a r}{\lambda v}\right) \right] \tag{1.83}$$

If the speckle 'size' is taken as the separation between the first two minima of the Bessel function, we have

$$d_{sp} = \frac{2.4\lambda v}{a} \tag{1.84}$$

since $J_1(x) = 0$ when $x = 1.22\pi$; this expression is the same as (1.81).

The size of the speckles in image-plane speckle is dependent on the aperture of the viewing lens – hence the use of the term 'subjective speckle'. This effect can be observed by comparing the size of the speckles when an object is viewed directly by eye and when an aperture smaller than the eye pupil is placed in front of the eye. In the latter case the speckle size will be seen to increase.

The maximum spatial frequency is now determined by the size of the viewing lens aperture and the distance from the lens to the viewing plane: it is given by

$$\frac{1}{f_{max}} = g_{min} \simeq \frac{\lambda v}{a} \tag{1.85}$$

1.8.3 *Determination of phase in coherent imaging*

Each of the techniques discussed in this book involves the observation of the light which is scattered from a rough surface illuminated with a coherent light source through a lens onto a plane which may be, but is not necessarily, the plane in which the object is in focus. In discussing these techniques it is necessary to be able to calculate the phase and the amplitude of the light in the observation plane. The approach which is outlined below is used in the book; this model avoids the use of complicated diffraction integrals while enabling the significant phase changes which arise in holographic and speckle interferometric experiments to be calculated.

Consider first the case where the object is in focus in the observation plane. The surface is illuminated by a wavefront diverging from the point S – see Figure 1.31. A point P in the surface illuminates an area in the image plane centred on Q; the diameter of this area is given by equation (1.81). The optical path from S to the point R via P is given by

$$l = SP + l_P + PO + OR \tag{1.86}$$

where SP and PO represent the mean distance from a region around P to S and O respectively, l_P is the optical path associated with the random variation of the surface height, and OR is the distance from O to R.

We now consider the case where the observation plane is located so that the lens is focussed on a point T (see Figure 1.32). The point Q in the viewing plane is illuminated by all the rays scattered from the surface which appear to diverge from T and which pass through the viewing lens aperture. Thus, the size of the area which illuminates a point in the

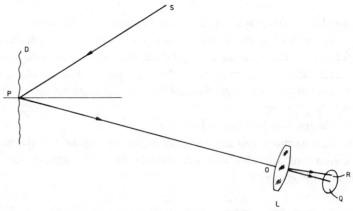

Fig. 1.31 The optical path of a ray: image-plane viewing

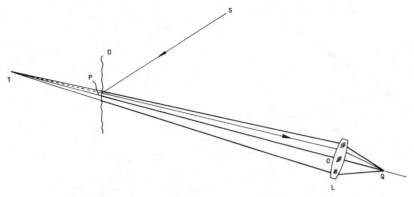

Fig. 1.32 The optical path of a ray: defocussed viewing

viewing plane is proportional to the viewing lens aperture diameter and inversely proportional to the distance between T and the viewing lens. Because the optical path from T to Q is the same for all ray directions, the optical path from S to Q via a point P is given by

$$l = SP + l_P + TQ - PT \tag{1.87}$$

When the viewing plane is the focal plane of the lens, T is effectively at infinity. All the rays illuminating a point Q in the viewing plane are parallel to MQ; the area illuminating a point in the viewing plane is now the projection of the viewing lens aperture on the object, so that unless viewing is at a very oblique angle, its diameter is approximately that of the viewing lens aperture.

The optical path from S to Q via a point P can be seen from Figure 1.33 to be given by

$$l = SP + l_P + MQ + PN \tag{1.88}$$

When the object is displaced, so that the point P is displaced by d to P′, the optical path from the source S to a point in the viewing plane via a given point in the object is altered. The change in phase associated with this change in optical path is the basis of holographic and speckle correlation techniques for measuring surface displacements, and its value is calculated here.

In each of the viewing arrangements of Figures 1.31, 1.32 and 1.33, the optical path from S to Q can be considered to be made up of three components;

    i) the path from S to P
    ii) the path l_p associated with the surface height variation and
    iii) the path from P to Q.

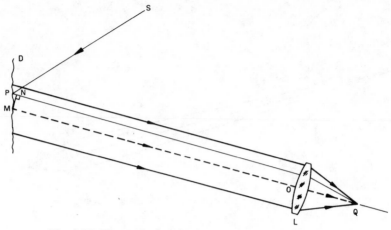

Fig. 1.33 The optical path of a ray: focal plane viewing

The path change Δl_1 due to (i) is given by:

$$\Delta l_1 = SP' - SP$$

– see Figure 1.34. For all practical arrangements $SP \gg |d|$ so that we have

$$\Delta l_1 = SP' - SP = d \cos \gamma_1 \tag{1.89}$$

If the direction of SP is given by a unit vector n_0, this may be written as

$$\Delta l_1 = n_0 \cdot d \tag{1.90}$$

(The assumption that $SP \gg |d|$ is equivalent to assuming that SP and SP' are approximately parallel i.e. $n_0' \simeq n_0$.)

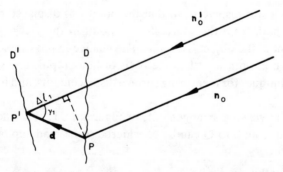

Fig. 1.34 The ray geometry for the calculation of the phase difference between the illuminating wavefronts introduced by a surface displacement d

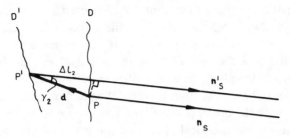

Fig. 1.35 The ray geometry for the calculation of the phase difference between the scattered wavefronts introduced by a surface displacement d

l_p can be assumed to be unchanged when the surface is displaced. The change in path Δl_2 associated with (iii) is given by

$$\Delta l_2 = d \cos \gamma_2 \qquad (1.91)$$

– see Figure 1.35 – where it is assumed that the direction in which the light is scattered from P to Q and from P′ to Q are approximately parallel. (These directions depend on which form of viewing – focussed or defocussed – is used). If this direction is given by n_s, we have

$$\Delta l_2 = -n_s \cdot d \qquad (1.92)$$

Thus, the total phase change due to the displacement is given by

$$\phi = \frac{2\pi}{\lambda}(n_0 - n_s) \cdot d \qquad (1.93)$$

1.9 Optical methods of Fourier transformation

It was seen in Section 1.4.2 that a general wavefront can be represented as the sum of a set of plane waves of different spatial frequencies (i.e. travelling at different angles). Several optical techniques enable the Fourier transform of the wavefront to be displayed as a variation of the intensity of light in a plane; the intensity at a given point is proportional to a given Fourier component, the coordinates of the point being linearly related to the direction cosines of the Fourier component. Two such techniques are discussed here; it will be seen in Chapter 3 that the Fourier transforming property of a lens is an essential part of speckle photography and photographic speckle correlation interferometry.

In the first of these, the wavefront is viewed in the focal plane of the viewing lens. (The illumination should be monochromatic and in the case of a transparency a plane wavefront should be used for illumination.) It was seen in Section 1.5.3 that when a plane wavefront is incident on a lens at an angle θ, it is brought to a focus at a point in the focal plane which is situated at a distance $f\theta$ from the principal axis of the lens where f is the focal length of the lens. In the case of a wavefront having direction cosines (l, m, n), the coordinates of the point of focus are given by $(fl, fm, 0)$, where the origin of the coordinate system is located at the point of intersection of the focal plane and the principal axis, and the angle between the plane wave and the optic axis is small i.e. $n \simeq 1$ (this will be accurate to 1% for $\cos^{-1} n = 8°$). Thus we see that the intensity at a point $(X, Y, 0)$ in the focal plane is proportional to the intensity of the Fourier component having direction cosines

$$l = \frac{X}{f}; \qquad m = \frac{Y}{f}$$

Viewing the wavefront in the focal plane of a lens is equivalent to viewing it at infinity without a lens; in speckle photography, the transparency is illuminated by an unexpanded laser beam and the Fourier transform of the transmitted wavefront which is proportional to that of the transparency can be observed by viewing the scattered light at a considerable distance ($\sim$a few metres) from the transparency without the use of a transforming lens.

The second technique which is frequently used in speckle interferometry to display the Fourier transform of a transparency, was first described by Burch and Tokarski (17). In this method, the transparency is illuminated by a converging wavefront (see Figure 1.36) and the light is viewed in the plane F which is the plane in which the converging wavefront would be focussed; the derivation in (17) can be extended to

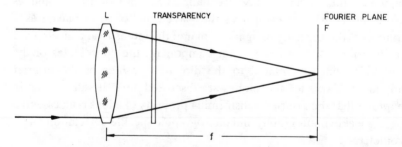

Fig. 1.36 An arrangement for viewing the Fourier spectrum of a transparency

two dimensions to give $U(X, Y)$, the amplitude at a point (X, Y) in the viewing plane, as

$$U(X, Y) = \exp{(i\alpha)} \int\int \rho(x, y)\exp\left\{-i\frac{2\pi}{\lambda z}[Xx + Yy]\right\} dx\ dy$$

$$= \exp{(i\alpha)}\mathscr{F}[\rho(x, y)]$$

(1.94)

where $\rho(x, y)$ describes the optical density of the transparency, and α is a phase factor. Thus, the intensity in the plane F is proportional to the Fourier transform of $\rho(x, y)$; the intensity at a point (X, Y) representing spatial frequency having components f_x, f_y given by

$$f_x = \frac{2\pi}{\lambda}\frac{X}{z}, \qquad f_y = \frac{2\pi}{\lambda}\frac{Y}{z}$$

(1.95)

where z is the distance between the lens and the viewing plane.

The plane in which the Fourier transform is mapped is known as the Fourier plane. The light near the origin represents the low frequency components, and further away the high frequency components are represented.

By placing an aperture in the Fourier plane it is possible to exclude a particular frequency component. A simple example of this is the use of a small circular hole (a pinhole) in the focal plane of a lens which is used to spread out a laser beam. By eliminating all but the very low frequencies, the aperture eliminates most of the noise arising from scattering by dust particles on the components preceding the aperture. This form of spatial filtering is frequently used in holographic and speckle interferometry to minimize optical noise – see, for example, Figure 4.8.

2

Holographic interferometry

2.1 Introduction: basic techniques and the general problem

It is the opinion of many scientists involved in the complex subject of surface deformation analysis that one of the most important techniques to emerge from the principle of holography (Section 1.7) is that of holographic interferometry. This enables the static and dynamic displacements of an optically rough surface to be measured interferometrically. First reports of the method appeared during the mid 1960s (for example, references 1–5) and were soon followed by numerous papers describing new general theories and applications (Section 2.9). One of the main reasons for such interest is that holographic interferometry clearly removes the most stringent limitation of classical interferometry (Section 1.5.5) i.e. that the object under investigation be optically smooth. Thus the advantages of interferometric measurement, for example, high sensitivity and non-contacting field view, can be extended to the investigation of numerous materials, components and systems previously outside the scope of optical study.

Let us first consider qualitatively how the holographic recording of a scattering surface can be used to detect the displacement of that surface. It has been shown in Section 1.7.3 that the developed hologram reconstructs a virtual image of the original object. If one views the precise superposition of the light from the reconstructed image and the real object through the hologram, then the interference of the two identical wavefronts results in a uniform field of view. This condition requires that the processed hologram be replaced in its original position to an accuracy of better than $\frac{1}{4}\lambda$ and also suffer minimum emulsion shrinkage. With care these conditions can be satisfied (Sections 6.4.2 and 6.5.2) to the extent that usually one broad, dark fringe (often referred to as a 'fluffed out fringe field') is obtained. When the real object undergoes a

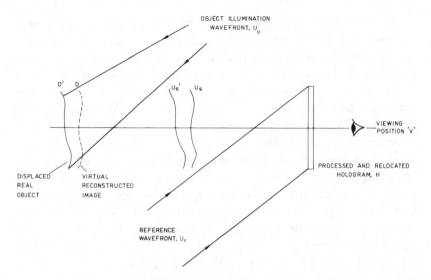

Fig. 2.1 Diagrammatic arrangement for the observation of 'real-time' or 'live' holographic interference fringes

small static displacement which is sufficient to produce a small variation in the relative phase of the two wavefronts, a fringe pattern can be observed. This is known as the 'real-time' or 'live' holographic interference fringe pattern. Figure 2.1 shows the basic physical arrangement. The hologram H reconstructs a virtual image of the object D. U_r is the reference wavefront and the object displaced to D' is illuminated by the original wavefront U_o. (Note that the paths travelled by object and reference beams are equalized using external optics not shown in Figure 2.1 so that they have travelled nominally the same distance at the plane of H. This ensures that they are mutually coherent. A practical layout is shown in Figure 2.11.) The wavefronts U_s and U_s' correspond respectively to the virtual image reconstructed wavefront and the wavefront scattered from the real, displaced object. These interfere to give a fringe pattern (for example, Figure 2.8), the form of which is dependent upon the geometry of the displacement and the direction of viewing. It is quite óften necessary for the observer at V to focus on a plane which does not lie in the plane of the object to see the fringes with maximum visibility. This plane of focus is usually referred to as the plane of fringe localization and is also a function of the displacement and viewing direction (Section 2.3.1). If the fringes have maximum visibility when the observer focusses on the object plane the fringes are said to be localized in the plane of the object.

An alternative technique known as 'frozen fringe' or 'double exposure' holographic interferometry may be used. This eliminates the need for precise plate relocation. In this method a hologram of the object in its undisplaced state is first recorded. The object is then displaced and a second holographic recording of D is superimposed upon the first. The hologram plate is held in the same position throughout. As a result of this the processed hologram simultaneously reconstructs the light fields corresponding to the object in the two positions. A single holographic interferogram is therefore stored by the hologram and can be observed in the absence of the original object. This is shown in Figure 2.2. Such a fringe field is commonly referred to as a 'frozen fringe pattern'. Although this method is less demanding experimentally than the live technique it does suffer from the disadvantage that only one displacement can be studied per holographic recording. This limitation does not have the same significance in the study of dynamic displacements in which frozen fringe holography plays an indispensable role.

Two of the most important techniques, time averaged holographic interferometry (1) (Section 2.6.1) and dual pulsed holographic interferometry (6) (Section 2.6.3) rely on the use of the frozen fringe method. The first of these is generally limited to the study of cyclic motions of small amplitude (typically in the range 1 to 10 λ) and requires that a hologram of the moving object be recorded for an exposure time well in excess of the period of oscillation. The frozen fringe pattern reconstructed from a hologram recorded in this way then represents the time average of the displacement distribution (for example, Figure 2.27). A

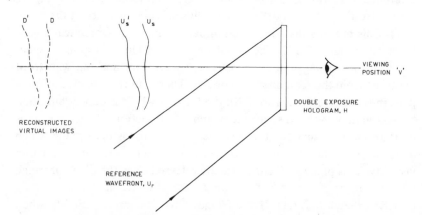

Fig. 2.2 Diagrammatic arrangement for the observation of 'frozen' or 'double-exposure' holographic interference fringes

wider range of dynamic displacements can·be investigated using the second technique. This relies on the use of a dual pulsed laser which is capable of delivering two short, high energy pulses of illumination in quick succession (7). Two holograms of a moving object can therefore be recorded on the same plate. The frozen fringe field reconstructed from such a double exposure hologram defines the distance moved by the object between the arrival of the two pulses. This movement must be of the order of a few wavelengths. However, the short pulse duration and separation enables general dynamic displacements, transients and small segments of large amplitude cyclic motions to be observed. Furthermore the short exposure times also enable such measurements to be made in the presence of random instabilities which would prevent the use of continuous wave (CW) laser holography (Sections 2.6.3, 3.4.3 and 4.7.4). This fact is demonstrated by Figure 2.30.

Holographic interferometry can also be used to detect optical path length variations in transparent media (for example R. F. Wuerker (8), 517–40). This is essentially a holographic version of the classical Mach–Zehnder interferometer described in Section 1.5.5. A basic arrangement is shown in Figure 2.3. Here the volume G of the gas or fluid under investigation is illuminated from the rear with the light scattered by a ground glass diffuser D. A hologram of G is recorded, processed and kinematically relocated in the manner described previously. The volume G is now subject to a small change in density which causes the optical path length of the light passing through it to change with respect to that recorded initially. It has already been noted that the refractive index is proportional to density, Section 1.5.5. A fringe pattern in which the

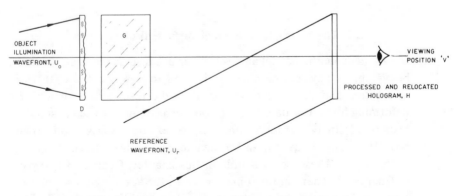

Fig. 2.3 Diagrammatic arrangement for the observation of real-time holographic fringes due to the variation in the optical path length in the transparent medium G

phase change is proportional to the integral of the steady state density variation through the volume G is therefore observed. The same type of fringe pattern can obviously be obtained using a Mach–Zehnder interferometer but the use of the hologram eliminates the need for matched, high precision, large aperture optical elements. The holographic arrangement also allows rapid density fluctuations to be studied when dual pulsed holographic interferometry is used (J. W. C. Gates and G. N. Hall (8), 115–26, see also Section 7.7).

The inherent advantages of holographic interferometry are somewhat counterbalanced by the problem of fringe pattern interpretation. This is often difficult because in the case of surface displacement fringe pattern analysis the fringe spacing is sensitive to all the individual components of the displacement present (Sections 2.2 and 2.4). The problem is relatively straightforward when the displacement has one or two components but becomes increasingly more difficult when rotations, deformations and rigid body translations occur simultaneously (Section 2.2.1). It is the purpose of the first part of this chapter to develop an approach to the general solution of this problem (Sections 2.3 and 2.4). This enables the fringe geometry and localization plane to be determined for various combinations of displacement. An important outcome of the analysis is that it provides the basis of a useful practical technique (Section 2.4.3). The second part of the chapter (Section 2.6) deals with the methods of dynamic displacement measurement outlined previously.

Fundamental to a large part of this work is an understanding of the way in which the geometry of a small general displacement is described mathematically. The first section therefore deals with this topic.

2.2 The description of small surface displacements

The mathematical analysis of surface displacement summarized below shows how tensor notation is used to describe the changes in orientation and shape that may occur as the result of the action of a deforming force. It is based on the approach commonly used in standard texts on elasticity and tensor analysis, for example Sokolnikoff (9) and Nye (10), and provides a basis for the understanding of the related topics in this book. These occur mainly in this chapter, Chapter 3 and parts of Chapter 7. There are two main reasons for adopting tensor and the associated repeated suffix notation. Firstly it enables complex displacements to be represented with a maximum economy of expression and secondly the same notation can be used to describe the parameters

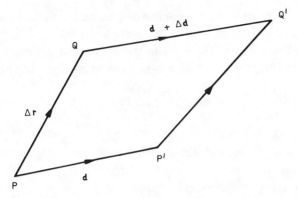

Fig. 2.4 The general displacement of a surface line element from PQ to P'Q'

governing fringe spacing and localization. In this way an overall continuity is maintained.

2.2.1 *The mathematical representation of small surface displacements*

In Figure 2.4, PQ represents a line element on the surface of a body. The body undergoes a general deformation after which the line is now at P'Q', where it has undergone a shift in position, orientation and length.

The position of P is represented by r, and that of Q by $r + \Delta r$. The position of P' is given by $r + d$ and that of Q' by $r + \Delta r + d + \Delta d$. Thus the displacement of Q can be represented by

$$L = d + \Delta d \tag{2.1}$$

If, for all points on the body, the displacement vector is given by $L = d$, i.e. $\Delta d = 0$, then the displacement represents a rigid body translation, i.e. the position of the body is altered but its shape and orientation remain unchanged. However if the displacement consists of a rotation and/or deformation, Δd is not zero. In this case L in equation (2.1) represents the displacement as a sum of the rigid body translation and a term describing rotation and deformation.

The term Δd may be written as a Taylor expansion as

$$\Delta d = d(r + \Delta r) - d(r)$$

$$= \frac{\partial d}{\partial x_1} \Delta x_1 + \frac{\partial d}{\partial x_2} \Delta x_2 + \frac{\partial d}{\partial x_3} \Delta x_3 + \cdots \tag{2.2}$$

where x_1, x_2, x_3 represent orthogonal axes, and Δx_1, Δx_2, Δx_3 are the components of Δd.

When the line element PQ is small, the higher order terms may be neglected. The components of Δd are given from equation (2.2) by

$$\Delta d_1 = \frac{\partial d_1}{\partial x_1}\Delta x_1 + \frac{\partial d_1}{\partial x_2}\Delta x_2 + \frac{\partial d_1}{\partial x_3}\Delta x_3 \qquad (2.3a)$$

$$\Delta d_2 = \frac{\partial d_2}{\partial x_1}\Delta x_1 + \frac{\partial d_2}{\partial x_2}\Delta x_2 + \frac{\partial d_2}{\partial x_3}\Delta x_3 \qquad (2.3b)$$

$$\Delta d_3 = \frac{\partial d_3}{\partial x_1}\Delta x_1 + \frac{\partial d_3}{\partial x_2}\Delta x_2 + \frac{\partial d_3}{\partial x_3}\Delta x_3 \qquad (2.3c)$$

where d_1, d_2 and d_3 are the components of d.

This can be written in the form

$$\Delta d_i = \sum_j \frac{\partial d_i}{\partial x_j}\Delta x_j \qquad i, j = 1, 2, 3$$

which can be further simplified to

$$\Delta d_i = \frac{\partial d_i}{\partial x_j}\Delta x_j \qquad i, j = 1, 2, 3 \qquad (2.4)$$

where the Einstein summation convention is used as follows: when a suffix occurs twice in the same term, summation with respect to all values of that suffix is understood.

The factors $\partial d_i/\partial x_j$ represent the components of a second rank tensor. In the present context the use of tensors can be regarded as a convenient notation to represent the displacement gradients of the object deformation; it is not appropriate here to discuss the properties of tensors in detail, suffice only to say that a tensor of the second rank describes a physical quantity which satisfies certain transformation relationships (9, 10), which are indeed satisfied by the components of equation (2.4). (A vector is a tensor of the first rank, and a scalar is a zero order tensor.) The tensor can be written in the form:

$$\begin{pmatrix} e_{11} & e_{12} & e_{13} \\ e_{21} & e_{22} & e_{23} \\ e_{31} & e_{32} & e_{33} \end{pmatrix} \qquad (2.5)$$

where

$$e_{ij} \equiv \frac{\partial d_i}{\partial x_j}$$

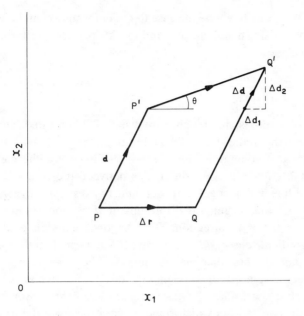

Fig. 2.5 The displacement of a line element PQ, parallel to the x_1-axis to P'Q' inclined at an angle θ to the x_1-axis

The physical significance of the components of the displacement gradient tensor can be seen more clearly if we consider a line element on a two-dimensional surface where the element is initially parallel to the x_1-axis, see Figure 2.5. The vector Δr has a component only in the x_1 direction, so that Δx_2, Δx_3 are zero. The components of Δd are given from equation (2.3) by

$$\Delta d_1 = \frac{\partial d_1}{\partial x_1}\Delta x_1$$

$$\Delta d_2 = \frac{\partial d_2}{\partial x_1}\Delta x_1$$

The term $\partial d_1/\partial x_1 = \Delta d_1/\Delta x_1$ represents the change in length per unit length along the x_1-direction – this is known as the normal or longitudinal strain.

It is seen that as well as being lengthened, PQ has been rotated by an angle θ where

$$\tan\theta = \frac{\Delta d_2}{\Delta x_1 + \Delta d_1} \simeq \frac{\Delta d_2}{\Delta x_1} \tag{2.6}$$

when $\Delta d_1 \ll \Delta x_1$, which will be the case in the deformation and observation conditions arising in holographic and speckle pattern interferometry. We then have

$$\lim_{\Delta x_1 \to 0} \frac{\Delta d_2}{\Delta x_1} = \frac{\partial d_2}{\partial x_1} = \theta \qquad (2.7)$$

Thus $\partial d_2 / \partial x_1$ represents the angle by which PQ is rotated. Similarly it can be shown that if the element PQ lying along the x_1-axis is deformed so that it has components in the x_1-, x_2-, x_3-directions, the elements $\partial d_2 / \partial x_1$ and $\partial d_3 / \partial x_1$ represent the angles between the components of the displaced line with respect to the x_2- and x_3-axes respectively.

Thus we see that a general displacement d can be resolved into components in the $x_{1,2,3}$-directions. The diagonal elements of the displacement gradient tensor, $\partial d_i / \partial x_i$ represent the normal strain for each component, and the non-diagonal elements, $\partial d_i / \partial x_j$ $i \neq j$, represent the angles between displaced jth component and the other two axes.

We will now show that the displacement gradient tensor can be separated into two tensors which separate rotation from deformation. To do this we will consider the form which the displacement gradient tensor has when a body undergoes a small rotation only. We consider two line elements PQ_1 and PQ_2 of equal length on a two-dimensional surface which are parallel to the x_1- and x_2-axes respectively. These are rotated by a small angle θ (see Figure 2.6). The components of the

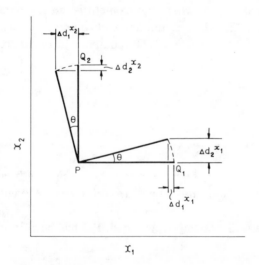

Fig. 2.6 The rotation of two orthogonal line elements PQ_1 and PQ_2 of equal length

displacement of PQ_1 are given by

$$\Delta d_1{}^{x_1} = \frac{\partial d_1}{\partial x_1} \Delta x_1$$

$$\Delta d_2{}^{x_1} = \frac{\partial d_2}{\partial x_1} \Delta x_1 = \theta \Delta x_1$$

It can be shown that for small θ, $\Delta d_1{}^{x_1} \ll \Delta d_2{}^{x_1}$, hence

$$\frac{\partial d_1}{\partial x_1} \simeq 0$$

The components of the displacement of PQ_2 are given by

$$\Delta d_1{}^{x_2} = \frac{\partial d_1}{\partial x_2} \Delta x_2 = \theta \Delta x_2$$

$$\Delta d_2{}^{x_2} = \frac{\partial d_2}{\partial x_2} \Delta x_2 \simeq 0$$

We see from Figure 2.6 that $\Delta d_1{}^{x_2} = -\Delta d_2{}^{x_1}$, so that

$$\frac{\partial d_1}{\partial x_2} = \frac{-\partial d_2}{\partial x_1} \tag{2.8}$$

Thus we see that when the body undergoes a small rotation, the components of the displacement gradient tensor satisfy the relationship

$$\frac{\partial d_i}{\partial x_j} = -\frac{\partial d_j}{\partial x_i} \qquad i, j = 1, 2 \tag{2.9}$$

A similar relationship can be derived for a three-dimensional surface:

$$\frac{\partial d_i}{\partial x_j} = \frac{-\partial d_j}{\partial x_i} \qquad i, j = 1, 2, 3 \tag{2.10}$$

The displacement gradients given by equation (2.4) represent both rotations and deformations of the body. The terms of this equation may be written as

$$\frac{\partial d_i}{\partial x_j} = \frac{1}{2}\left(\frac{\partial d_i}{\partial x_j} + \frac{\partial d_j}{\partial x_i}\right) + \frac{1}{2}\left(\frac{\partial d_i}{\partial x_j} - \frac{\partial d_j}{\partial x_i}\right)$$

$$\equiv \varepsilon_{ij} + \omega_{ij} \tag{2.11}$$

It can be seen that the ω_{ij} terms satisfy equation (2.10); if the ε_{ij} terms are zero, then the body undergoes rotation only. In the general case, the ω_{ij} terms represent that part of the displacement which constitutes rotation of the body and the ε_{ij} terms represent the deformation; the latter are the components of a tensor known as the strain tensor.

2.3 An analysis of holographic fringe formation for static surface displacements

In this section an expression for the relative phase of the wavefronts forming the holographic fringe pattern is determined. This result is used to derive equations which enable planes of fringe localization to be calculated. (Static surface displacements and collimated object illumination are assumed throughout this section and Section 2.4.)

2.3.1 *Conditions for holographic fringe observation and localization*

Holographic fringes form as the result of the interference between the light scattered from two identical surfaces D and D' placed at slightly different positions in space. This is shown in Figures 2.1 and 2.2. (D in Figure 2.1 and D, D' in Figure 2.2 are not real surfaces but in terms of the light field present may be referred to as such.) The wavefronts are made to interfere by using a viewing lens to bring them to focus in a common plane. This lens may, for example, be the eye of the observer or the camera lens if it is required to photograph the fringes. The two basic problems are to explain how these fringes form and why the plane of focus of the viewing lens for which the fringes are observed with maximum visibility varies as a function of displacement and viewing direction.

The light at a point in the viewing plane is scattered from an area in the object whose size is determined by the location of that plane; this area will be referred to as the resolution area. When the viewing plane corresponds to the focal plane of the viewing lens, the diameter of the resolution area will be approximately that of the viewing lens aperture (up to several centimetres) while when the viewing plane is the plane in which the object is focussed, the diameter is approximately that of the Airy resolution diameter or disc (see Section 1.8.3) which will be of the order of tens of microns.

When the surface is displaced the phase of the light scattered from each point on the surface to each point in the viewing plane is altered. The phase ψ of the light scattered from a given point in the undisplaced object to a given point in the viewing plane is related to the phase ψ' of the light scattered from the same point in the displaced object to the same point in the viewing plane (see equation 1.93) by

$$\psi' = \psi + (2\pi/\lambda)(n_o - n_s) \cdot d \qquad (2.12)$$

where d is the displacement of the object point and n_0 and n_s are the directions of illumination and viewing of that point.

In order to find how the intensity of the light in the viewing plane changes as a result of the object displacement, we first give an expression for the total amplitude of the light scattered to a point in the viewing plane from its resolution area; the surface will be considered to be made up of a set of point scatterers of random height which scatter uniformly in all directions. (A surface is more exactly represented by a continuum but the use of the discrete model here enables results to be derived relatively simply which are equivalent to those obtained using an exact description and a necessarily more difficult analysis.)

The amplitude of the light arriving at a point in the viewing plane before displacement can be represented by

$$U_1 = u_o \sum_{p=1}^{n} \exp i\psi_p \qquad (2.13)$$

where u_o is the amplitude of the light scattered by an individual scattering point, ψ_p is the phase of the light scattered by the pth point, and the resolution element contains n points.

The amplitude of the light at the same point after the object has been displaced is

$$U_2 = u_o \sum_{p=1}^{n} \exp i(\psi_p + \phi_p)$$

where

$$\phi_p = (2\pi/\lambda)[(n_o - n_s) \cdot d_p] \qquad (2.14)$$

is the phase change associated with the displacement of the pth scattering point.

When the two light fields are superimposed, the resultant intensity is given by

$$I = (U_1 + U_2)(U_1^* + U_2^*)$$

$$= u_o^2 \sum_{p} \sum_{q} \exp i(\psi_p - \psi_q) + \exp i(\psi_p' - \psi_q')$$

$$+ 2 \cos(\psi_p + \phi_p - \psi_q) \qquad (2.15)$$

Because the phase terms ψ_p and ψ_q contain random phase terms due to random variation of the surface height (see Section 1.3.2), the intensity

I varies randomly across the viewing plane. The mean value of *I* can be found by separating the summation of equation (2.15) into terms with $p = q$, and terms with $p \neq q$, giving

$$I = 2u_o^2 \left(n + \sum_p \cos \phi_p \right)$$

$$+ u_o^2 \left[\sum_p \sum_{q \neq p} \{ \exp i(\psi_p - \psi_q) + \exp i(\psi_p' - \psi_q') \right.$$

$$\left. + 2 \cos (\psi_p + \phi_p - \psi_q) \} \right] \tag{2.16}$$

The cross-terms vanish when this expression is averaged over many viewing plane points because terms of the form $\langle \sum_p \cos \chi_p \rangle = \langle \sum \sin \chi_p \rangle = 0$ when χ_p varies randomly.

In general the value of ϕ_p also varies randomly so that the mean value of *I*, the intensity in the viewing plane is given by

$$\langle I \rangle = 2u_o^2 n \tag{2.17}$$

If, however, the value of ϕ_p is constant across the resolution element, the value of *I* at a given point is given by

$$I = 2u_o^2 n (1 + \cos \phi_p) + \text{cross-terms} \tag{2.18}$$

If equation (2.18) is averaged over many speckles along a line in the viewing plane for which ϕ_p is constant both across the resolution elements and for all the points along the line, then the mean value of *I* is given by

$$\langle I \rangle = 2u_o^2 n [1 + \cos \phi_p] \tag{2.19}$$

It is seen that the mean value of the intensity varies with the value of ϕ_p, having a maximum value of $4u_o^2 n$ when $\phi_p = 2n\pi$ and a minimum value of zero when $\phi_p = (2n + 1)\pi$; this variation in the mean intensity of the speckle pattern is the holographic fringe pattern.

The value of ϕ_p across the resolution element can be seen to be constant only when the following conditions are satisfied:

(1) the illumination is a plane wavefront so that n_o is constant;
(2) the viewing plane is in the focal plane of the lens so that n_s is constant; and
(3) the body undergoes a rigid body translation so that d is constant.

The first and third of these should be obvious; the second follows because when the recording plane is located in the focal plane of the lens, the light arriving at a given point in that plane is scattered in the

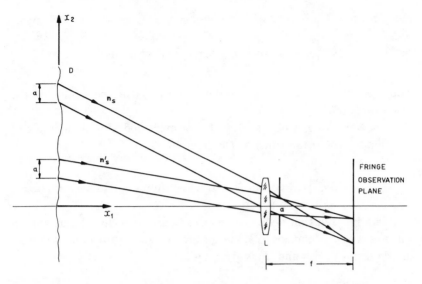

Fig. 2.7 Resolution elements in the object surface D as observed with the viewing lens, L, focussed on infinity. Note that the elements have the same size as the lens aperture a

same direction from all the points in the resolution element (see Figure 2.7). Thus it is seen that ϕ_p is the same for all the points in the resolution element illuminating a given point in the viewing plane.

It can also be seen from Figure 2.7 that the direction of n_s varies as the position in the viewing plane is altered, and this variation in ϕ_p across the viewing plane gives rise to a holographic fringe pattern. The relationship between the form and spacing of the fringes and the magnitude and direction of the rigid body displacement is discussed in Section 2.5.1. Since the lens is focussed on infinity, these fringes are said to be localized at infinity.

The visibility of holographic fringes can be defined by an expression similar to equation (1.41) as

$$\frac{\langle I \rangle_{\max} - \langle I \rangle_{\min}}{\langle I \rangle_{\max} + \langle I \rangle_{\min}} \tag{2.20}$$

and it can be seen that the fringes due to rigid body translations have a visibility of unity.

In general, the value of ϕ_p is not the same for all the points in the resolution area; holographic fringes may nonetheless be observed by a suitable choice of viewing plane. To see this, equation (2.16) is written in the form

$$I = 2u_o{}^2 \left[n + \sum_{p=1}^{n} \cos \left(\phi_0 + \Delta \phi_p \right) \right] + \text{cross-terms} \qquad (2.21)$$

where $\phi_p = \phi_0 + \Delta \phi_p$, and ϕ_0 is the value of ϕ at the centre of the resolution area.

If the value of I is now averaged over many speckles along a line of constant ϕ_0, the mean value of I is given by

$$\langle I \rangle = 2u_o{}^2 \left[n + \cos \phi_0 \sum_{p=1}^{m} \cos \Delta \phi_p + \sin \phi_0 \sum_{p=1}^{n} \sin \Delta \phi_p \right] \qquad (2.22)$$

It can be seen that if the values of $\Delta \phi_p$ are such that the maximum value of $\Delta \phi_p \ll \pi$, the $\cos (\Delta \phi_p)$ terms are approximately equal to unity, the sin terms vanish and $\langle I \rangle$ is given by

$$\langle I \rangle \approx 2u_o{}^2 n [1 + \cos \phi_0] \qquad (2.23)$$

so that a holographic fringe pattern is observed which represents the variation of ϕ_0.

If the object is assumed to lie in the $x_2 x_3$-plane, and the phase change associated with the light scattered from a point in the object which is initially located at (r), to a given point in the viewing plane, is given by $\phi(r)$, the variation of $\phi(r)$ across the resolution area can be represented by a Taylor expansion as

$$\phi(r) = \phi_0 + \left(\frac{\partial \phi}{\partial x_2} \right) (x_2 - x_{0,2}) + \frac{\partial \phi}{\partial x_3} (x_3 - x_{0,3}) + \cdots \qquad (2.24)$$

where $\phi(r_0) \equiv \phi_0$, and the coordinates of r_0 are $x_{0,2}$, $x_{0,3}$.

The maximum value of $\Delta \phi = \phi(r) - \phi_0$ is obtained at the extreme points in the resolution element, and it can be seen that the variation in $\phi(r)$ across the element is minimized if

$$\Delta \phi_{\text{max}} = \left(\frac{\partial \phi}{\partial x_i} \right) \Delta x_i \rightarrow \text{minimum} \qquad i = 2, 3 \qquad (2.25)$$

where Δx_i represents the size of the resolution element in the x_i directions.

It can be seen from equations (2.20), (2.21) and (2.25) that unless $(\partial \phi / \partial x_i) \Delta x_i$ is zero, the visibility of the fringes is less than unity, since $\langle I \rangle_{\text{min}}$ is not zero. However, maximum visibility fringes are obtained when equation (2.25) is satisfied, and the fringes are then said to be localized in the plane in which this occurs.

Since $\phi = 2\pi/\lambda\,(n_o - n_s)\cdot d$, equation (2.25) can be written as

$$\left[\frac{\partial}{\partial x_i}(n_o - n_s)\cdot d + (n_o - n_s)\cdot\frac{\partial d}{\partial x_i}\right]\Delta x_i \to \text{minimum} \qquad (2.26)$$

and this is the general condition which must be satisfied for holographic fringe localization.

Consider now the case where the displacement of the body consists purely of displacement gradients (i.e. $d = 0$); equation (2.26) reduces to

$$\left[(n_o - n_s)\cdot\frac{\partial d}{\partial x_i}\right]\Delta x_i \to \text{minimum} \qquad (2.27)$$

and this condition can be satisfied only by having a minimum value of Δx_i; this occurs when the viewing plane is located in the image plane of the object. Thus holographic fringes which result from a displacement gradient only are observed with maximum visibility when the viewing lens is focussed on the object; these fringes are said to be localized on the object surface. The contrast of such fringes is generally very good because the value of Δx_i is sufficiently small for $(\partial\phi/\partial x_i)\Delta x_i$ to be very much less than π; an example is shown in Figure 2.8 where the surface has undergone an out-of-plane rotation $d_{1,2}$. (In Sections 2.5.2 and 2.5.3 the relationship between the form and spacing of the fringes observed and the magnitude and direction of the displacement gradient is discussed.)

When the deformation of the object consists of a combination of rigid body translation and displacement gradients (which of course include rotations), the plane in which equation (2.26) is satisfied represents a compromise between the requirement of the first term that the viewing plane be located in the focal plane of the lens (i.e. $\partial n_s/\partial x = 0$) and that of the second term that the viewing plane be located in the image plane (i.e. Δx_i minimum). A particular example is discussed in Section 2.5.4.

When studying surface deformation using holographic interferometry it is desirable, however, to keep rigid body translations to a minimum so that the fringes are localized on or very near to the object surface (see also Section 2.8.1).

2.4 The measurement of surface displacements from holographic interferograms

In this section we will look at how the holographic fringe pattern can be used to measure the magnitude and direction of the displacement and the displacement gradient at a point in the object.

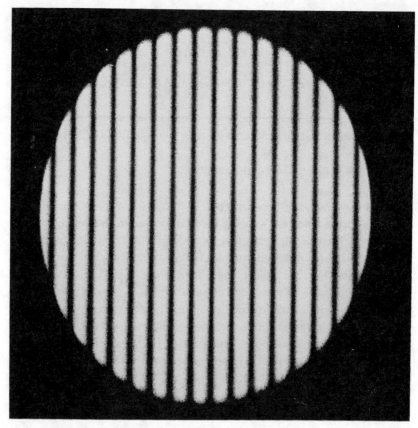

Fig. 2.8 Holographic fringes localized in the plane of the object due to an out-of-plane rotation $d_{1,2}$

In Section 2.4.1 the variation of the fringe order number N (see Section 2.4.1) with the magnitude and direction of the object displacement will be described in terms of a quantity known as the sensitivity vector. It is then shown (Section 2.4.2) how the fringe spacing in an area of the viewing plane is related to the displacement gradient and rigid body translation of the corresponding region in the object. This approach has the advantage that it enables the components of displacement to be more readily identified from the geometry of the fringe pattern than in the case when only the fringe order number is considered. Furthermore if the value of N is found for a single point, the variation of N over the whole surface can be found by measuring the fringe spacing over the surface, provided the signs of the gradients of N are known. The evaluation of the total displacement function is thereby considerably simplified.

A holographic arrangement which further simplifies the fringe pattern interpretation is described in Section 2.4.3. Finally ways in which the sign of N and $\partial N/\partial x$ may be determined are discussed in Section 2.4.4.

2.4.1 *The sensitivity vector*

Consider Figure 2.9. The object D is illuminated by a plane wave incident in the direction $\boldsymbol{n}_0$. A hologram of the object is recorded at H and the light scattered from the undisplaced and displaced object is viewed through the hologram in a plane which satisfies the fringe localization condition (i.e. the viewing lens is focussed upon the plane of fringe localization). The fringe order number N at a point in the viewing plane is defined as

$$N = \frac{\phi_0}{2\pi} = \frac{1}{\lambda}(\boldsymbol{n}_0 - \boldsymbol{n}_s) \cdot \boldsymbol{d} \qquad (2.28)$$

where ϕ_0 is the phase change associated with the displacement at that point. It should be noted that N is a continuous function, not an integer.

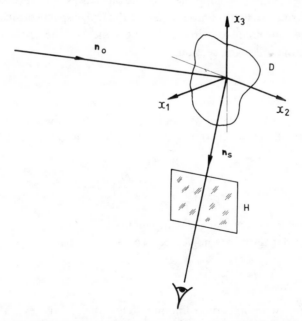

Fig. 2.9 The geometry for the observation of holographic fringes. The quasi-plane object D lies in the x_2x_3-plane and is observed through the hologram H in the direction $\boldsymbol{n}_s$ where $\boldsymbol{n}_0$ is the object illumination direction

We can write

$$N\lambda = (n_i^\circ - n_i^s)d_i \qquad i = 1, 2, 3 \qquad (2.29)$$

where n_i°, n_i^s are the components of the unit vectors n_o, n_s in the ith direction. Equation (2.29) can be written in the form

$$N\lambda = C_i d_i \qquad (2.30)$$

where $C_i = (n_i^\circ - n_i^s)$ is the sensitivity vector.

We see from equation (2.30) that a single measurement of N at a given point in the fringe pattern does not enable the displacement d at a point in the object to be determined, since it contains three independent variables d_1, d_2, d_3. For a given point in the object it is necessary to measure three independent values of N for three different viewing directions (i.e. different C_i) if the three components of the complete vector displacement are to be determined. The signs of N must also be found (see Section 2.4.4). Three independent equations of this form are then obtained from which the components of d can be calculated. This approach can often be quite difficult in practice (Section 2.7.1). Fringe pattern interpretation is simplified if one also uses the relationship between the fringe spacing components (i.e. the fringe spacings measured parallel to the coordinate axes in the viewing plane) and the displacement as derived in the following section.

2.4.2 *The fringe spacing equation*

If the fringe order number at a point in the pattern corresponding to a point in the object at position vector r_o is $N(r_o)$ and that at another point is $N(r)$, we have

$$N(r) = N(r_0) + \left(\frac{\partial N}{\partial x_i}\right)(x_i - x_i^0) + \left(\frac{\partial^2 N}{\partial x_i^2}\right)\frac{(x_i - x_i^0)^2}{2!} + \cdots$$

where the $(x_i - x_i^0)$ represent the components of $(r - r_0)$.

If higher order terms are neglected, we have

$$\Delta N = \left(\frac{\partial N}{\partial x_i}\right)(x_i - x_i^0) \qquad (2.31)$$

If r and r_o are two points on a line parallel to the x_j axis separated by Δx_j, when Δx_j is the separation of the fringes along that line we have

$$\left(\frac{\partial N}{\partial x_j}\right)\Delta x_j = 1 \qquad (2.32)$$

Differentiation of equation (2.30) gives

$$\lambda\left(\frac{\partial N}{\partial x_j}\right) = \left(\frac{\partial C_i}{\partial x_j}\right)d_i + C_i\left(\frac{\partial d_i}{\partial x_j}\right)$$

and we can write, using equation (2.32)

$$\frac{\lambda}{\Delta x_j} = \left(\frac{\partial C_i}{\partial x_j}\right)d_i + C_i\left(\frac{\partial d_i}{\partial x_j}\right) \tag{2.33}$$

Thus, the spacing of the fringes Δx_j is related to the components of the rigid body translation d_i, and those of the displacement gradients $\partial d_i/\partial x_j$ (equations 2.10 and 2.11).

Equation (2.33) may be written in the form

$$\frac{\lambda}{\Delta x_j} = C_{i,j}d_i + C_i d_{i,j} \tag{2.34}$$

where

$$C_{i,j} \equiv \frac{\partial C_i}{\partial x_j} \quad \text{and} \quad d_{i,j} \equiv \frac{\partial d_i}{\partial x_j}$$

The object surface can be assumed to lie in the x_2x_3-plane without loss of generality when the surface is plane.

The expansion of the repeated suffix notation enables the fringe spacing equations to be written in full as

$$\frac{\lambda}{\Delta x_2} = C_{1,2}d_1 + C_1 d_{1,2} + C_{2,2}d_2 + C_2 d_{2,2}$$

$$+ C_{3,2}d_3 + C_3 d_{3,2} \tag{2.35a}$$

$$\frac{\lambda}{\Delta x_3} = C_{1,3}d_1 + C_1 d_{1,3} + C_{2,3}d_2 + C_2 d_{2,3}$$

$$+ C_{3,3}d_3 + C_3 d_{3,3} \tag{2.35b}$$

These equations, in conjunction with equations (2.30) enable the components of the displacement d_i, and the components of the displacement gradient $d_{i,j}$ to be determined if measurements in three independent viewing directions are made; i.e. the nine measurements [$3 \times (N, \Delta x_2, \Delta x_3)$] enable the nine quantities d_i, $d_{i,j}$ to be found. It is seen that each viewing direction involves the calculation of nine independent coefficients C_i and $C_{i,j}$ which depend on the viewing geometry when the object surface is assumed to lie in the x_2x_3-plane. (Twelve coefficients are required in the general case.)

The spacing of the fringes will of course be measured in the image plane, whereas the Δx_j refer to the object plane coordinate. If the object is viewed in the normal direction the fringe spacing in the image plane $(\Delta x_j)_{im}$ is related to the fringe spacing in the object plane $(\Delta x_j)_{obj}$ by

$$(\Delta x_j)_{im} = \frac{S_2}{S_1}(\Delta x_j)_{obj} \qquad (2.36)$$

where S_1, S_2 are the lens to object and lens to image plane distances respectively. This applies regardless of where the lens is focussed since the (Δx_j) in equation (2.33) are always referred to the object surface.

If the fringes are viewed at an angle θ to the surface-normal from a point in the $x_1 x_2$-plane, as shown in Figure 2.10 it can be seen that the $(\Delta x_2)_{im}$ is now given by

$$(\Delta x_2)_{im} = \frac{S_2}{S_1}(\Delta x_2)_{obj} \cos \theta \qquad (2.37)$$

since the normal to the viewing plane will also be inclined at an angle θ to the surface-normal. The effect must always be taken into account when viewing the fringes in a non-normal direction.

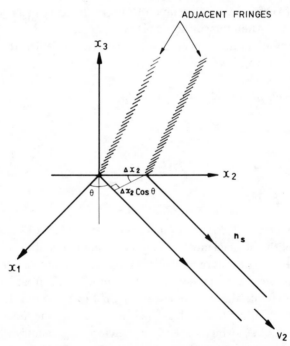

Fig. 2.10 The resolved component of the Δx_2 fringe spacing, Δx_2 $\cos \theta$, as observed at V_2

Finally it should be noted that a measurement of the fringe spacing does not indicate the sign of $(\partial N/\partial x_j)$, only its magnitude. The sign may be determined from a calibration of the fringe motion that is seen to occur as the displacement takes place. This technique is described in Section 2.4.4.

2.4.3 *The application of the sensitivity vector and fringe spacing equations*

It was seen in Sections 2.4.1 and 2.4.2 that measurements must be made for three independent viewing directions in order to determine the components of a displacement (d_i) and displacement gradient $(d_{i,k})$

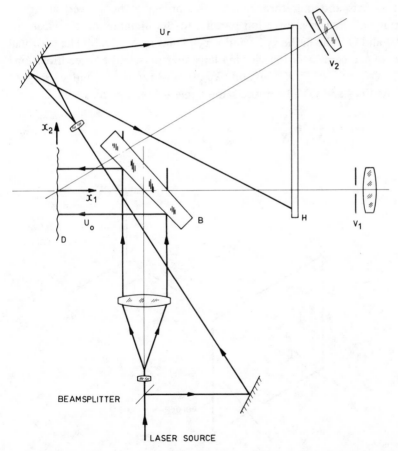

Fig. 2.11 A practical arrangement for a holographic interferometer. The aperture of the hologram is sufficient to accommodate the V_1, V_2 and V_3 viewing directions as shown in Figure 2.12

at a point in the interferogram. When the axes are defined such that the object surface lies in the $x_2 x_3$-plane, nine coefficients must be calculated for each measurement in the three 'general' viewing directions. In this section an arrangement is described which reduces the total number of coefficients from 27 to 13 and hence simplifies the fringe pattern analysis.

The object D is located in the $x_2 x_3$-plane as shown in Figure 2.11. It is illuminated by a plane wavefront U_0 which is reflected from a beamsplitter B. A hologram of D is recorded in the plane H which is perpendicular to the x_1 axis. B and H are large enough to enable three viewing directions OV_1, OV_2 and OV_3, Figure 2.12, to be accommodated by the single hologram recording. V_1, V_2 and V_3 represent the centres of the viewing lens apertures for the three directions. OV_1 lies along the x_1-axis and therefore enables the object to be viewed along its surface-normal in a direction parallel to the illumination direction $\boldsymbol{n}_0$. OV_2 and OV_3 lie in the $x_1 x_2$- and $x_1 x_3$-planes respectively. The coordinates of the centres of the viewing lens apertures in the three directions are $(X_{11}, 0, 0)$, $(X_{21}, X_{22}, 0)$ and $(X_{31}, 0, X_{33})$ for V_1, V_2 and V_3 respectively. OV_2 and OV_3 lie at the same angle θ to the x_1-axis, i.e.

$$\tan \theta = \frac{X_{22}}{X_{21}} = \frac{X_{33}}{X_{31}} \tag{2.38a}$$

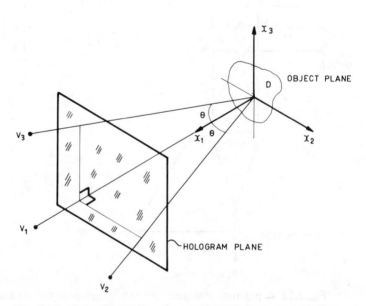

Fig. 2.12 Viewing positions for the interferometer shown in Figure 2.11

OV_1, OV_2 and OV_3 are of the same length and hence

$$\left.\begin{array}{l} X_{21} = X_{31} \\[6pt] X_{22} = X_{33} \\[6pt] X_{11} = (X_{21}{}^2 + X_{22}{}^2)^{\frac{1}{2}} = R \end{array}\right\} \qquad (2.38b)$$

The distances OV_1, OV_2 and OV_3 are assumed to be large with respect to the size of the object. The coefficients C_i, $C_{i,j}$ (equation 2.34) must be calculated for each viewing position. This requires that the following equations be expressed in terms of the object and viewing position coordinates:

$$(n_o - n_1{}^s) \cdot d = N_1 \lambda \qquad (2.39a)$$

$$(n_o - n_2{}^s) \cdot d = N_2 \lambda \qquad (2.39b)$$

$$(n_o - n_3{}^s) \cdot d = N_3 \lambda \qquad (2.39c)$$

where n_o is the illumination direction vector $n_1{}^s$, $n_2{}^s$ and $n_3{}^s$ represent the directions of the light scattered from a given point in the object to the centre of the viewing lens aperture at V_1, V_2, V_3 respectively (Figures 2.11 and 2.12) and N_1, N_2, N_3 are the fringe order numbers for the three viewing positions.

When these equations are evaluated and differentiated with respect to the object-plane coordinates (x_2, x_3), the results given in Table 2.1 are obtained. Here the following extension of the initial notation (equation 2.30) has been made: C_i coefficients for the OV_k, $(k = 1, 2, 3)$ viewing directions are now written as C_{ki}. For example, the coefficients become C_{21}, C_{22} and C_{23} for the OV_2 viewing direction. Similarly the differential coefficients $C_{i,j}$ are written $C_{ki,j}$ for each of the $k = 1, 2, 3$ viewing directions so that for the V_2 viewing position they will consist of

$$C_{21,2}, \quad C_{21,3}, \quad C_{22,2}$$

$$C_{22,3}, \quad C_{23,2}, \quad C_{23,3}$$

Also, in Table 2.1, $R = OV_k$ $(k = 1, 2, 3)$.

In order to see how the results in Table 2.1 are obtained we will consider the derivation of the C_{22}, $C_{22,2}$, C_{12} and $C_{12,2}$ coefficients. For the V_2 viewing position we have, from equation $(2.39b)$ and Figure 2.13 that for a point in the object having coordinate $(0, x_2, 0)$ that has been displaced a distance d_2,

$$N_2 \lambda = d_2 \sin [\theta(x_2)]$$

Table 2.1. C_{ki} and $C_{ki,j}$ coefficient for OV_1, OV_2, OV_3 viewing directions

OV_1 viewing direction

$C_{11} = 2$	$C_{12} \simeq 0$	$C_{13} \simeq 0$
$C_{11,2} = 0$	$C_{12,2} = \dfrac{1}{R}$	$C_{13,2} = 0$
$C_{11,3} = 0$	$C_{12,3} = 0$	$C_{13,3} = \dfrac{1}{R}$

OV_2 viewing direction

$C_{21} = (1 + \cos \theta)$	$C_{22} = \sin \theta$	$C_{23} \simeq 0$
$C_{21,2} = \dfrac{\tan \theta}{X_{21}(1 + \tan^2 \theta)^{\frac{3}{2}}}$	$C_{22,2} = -\dfrac{\cos^2 \theta}{R}$	$C_{23,2} = 0$
$C_{21,3} = 0$	$C_{22,3} = 0$	$C_{23,3} = \dfrac{1}{R}$

OV_3 viewing direction

$C_{31} = (1 + \cos \theta)$	$C_{32} \simeq 0$	$C_{33} = \sin \theta$
$C_{31,2} = 0$	$C_{32,2} = \dfrac{1}{R}$	$C_{33,2} = 0$
$C_{31,3} = \dfrac{\tan \theta}{X_{31}(1 + \tan^2 \theta)^{\frac{3}{2}}}$	$C_{32,3} = 0$	$C_{33,3} = -\dfrac{\cos^2 \theta}{R}$

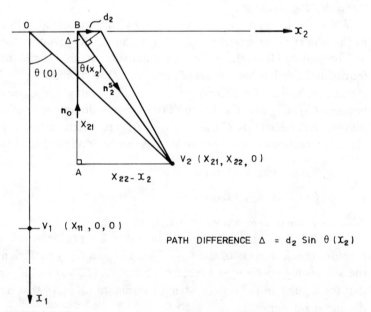

Fig. 2.13 The geometry for the calculation of the C_{22} and $C_{22,2}$ coefficients. (The object lies in the $x_2 x_3$-plane)

and hence by comparison with equations (2.30) and (2.39),

$$C_{22} = \sin \theta(x_2) \tag{2.40}$$

where $\theta(x_2)$ is as shown in Figure 2.13 and $x_2 < X_{22}$. Equation (2.40) may be written as

$$C_{22} = \frac{AV_2}{BV_2} = \frac{X_{22} - x_2}{[(X_{22} - x_2)^2 + X_{21}{}^2]^{\frac{1}{2}}}$$

and

$$C_{22,2} = \frac{\partial}{\partial x_2}(C_{22}) = \frac{\partial}{\partial x_2}\left\{\frac{X_{22} - x_2}{[(X_{22} - x_2)^2 + X_{21}{}^2]^{\frac{1}{2}}}\right\}$$

This is evaluated to give

$$C_{22,2} = -\frac{X_{21}{}^2}{[(X_{22} - x_2)^2 + X_{21}{}^2]^{\frac{3}{2}}}$$

When $x_2 \ll X_{22}$, we can write

$$C_{22} = \sin \theta \tag{2.41}$$

and

$$C_{22,2} = -\frac{\cos^2 \theta}{R} \tag{2.42}$$

where $\theta \equiv \theta(0)$. The maximum value of x_2 equals $\frac{1}{2}W$ where W is the width of the object and the condition

$$\tfrac{1}{2}W \ll X_{22}$$

will be satisfied if

$$\tan \theta \gg \frac{W}{2R}$$

As the viewing angle approaches the normal direction, this condition is no longer satisfied, and when $X_{22} < x_2$, we have

$$C_{22} = -\frac{(X_{22} - x_2)}{[(X_{22} - x_2)^2 + X_{21}{}^2]^{\frac{1}{2}}}$$

so that at V_1 (i.e. $X_{22} = 0$)

$$C_{12} = \frac{x_2}{(x_2{}^2 + X_{21}{}^2)^{\frac{1}{2}}} \tag{2.43}$$

and

$$C_{12,2} = \frac{1}{R} \qquad (2.44)$$

It can be seen that C_{11} is considerably greater than C_{12} and for this reason C_{12} is written as being approximately equal to zero. In physical terms this result means that the normal viewing direction is typically two orders of magnitude more sensitive to out-of-plane than in-plane displacement gradients. (A sensitivity to in-plane displacement gradients is introduced as the viewing direction departs from the normal.) By comparison equations (2.42) and (2.43) indicate that $C_{22,3}$ and $C_{12,3}$ are precisely equal to zero and are therefore written as such in Table 2.1.

A similar analysis to that outlined above is used to derive the remaining coefficients in Table 2.1. The physical application of these results is discussed in Section 2.5.

2.4.4 *Measurement of the fringe order number N, and the sign of $\partial N / \partial x_j$*

It was seen in Section 2.4.3 that the magnitude and direction of the displacement d of an object may be determined from a holographic fringe pattern if values for the fringe order number N in three viewing directions are known. N is given by the general equation $N\lambda = (n_0 - n_s) \cdot d$ (see Section 2.4.3). Clearly a single static fringe pattern contains no information about the absolute value of N, and even relative values of N cannot be unambiguously determined.

It will be shown, however, that the value of N can be found in live holographic interferometry by observing the motion of the fringes as the force which gives rise to the displacement is applied*. When the function which describes the variation in N is simple, this procedure is fairly straightforward but if it has several maxima and minima whose positions perhaps alter as the load is applied, the measurement of N can be quite difficult.

In double-exposure holographic interferometry N can be measured only if the form of the variation of N is known and is fairly simple, and a point of known N can be identified.

Consider first some fairly simple N distributions. The continuous curve A in Figure 2.14(*a*) represents a variation of N with respect to

---

* Here it is important to note that unless the displacement is applied smoothly the fringe motion will not be resolvable. In this context hydraulic load application has been found to be ideal (see also Section 7.4.1).

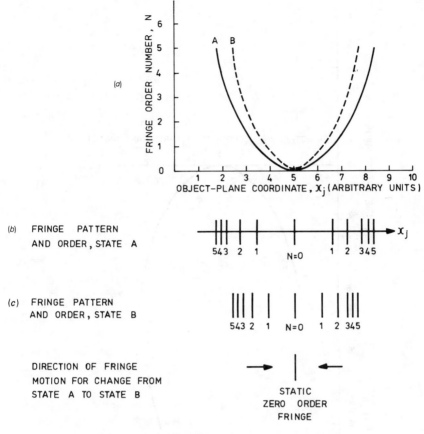

Fig. 2.14 (*a*) Fringe order number variations N with respect to object-plane coordinate x_j and fringe pattern corresponding to the variation A. (*b*) The fringe pattern corresponding to the fringe order number variation B in Figure (*a*). The direction of fringe motion that occurs in passing from state A to state B is indicated

x_j ($j = 2, 3$) for the application of a given load. Here x_j is the object plane coordinate and Figure 2.14(*b*) also shows the fringe pattern which results from that distribution. The dotted curve B represents the distribution of N for the same object for an increased load. The fringes now have the form shown in Figure 2.14(*c*) and it is seen that the fringes move towards the point where $N = 0$ (zero fringe order number) as the object moves from A to B.

It can be readily seen that the distributions given by Figure 2.15(*a*)–(*c*) give fringe patterns whose form and motion are similar to those indicated

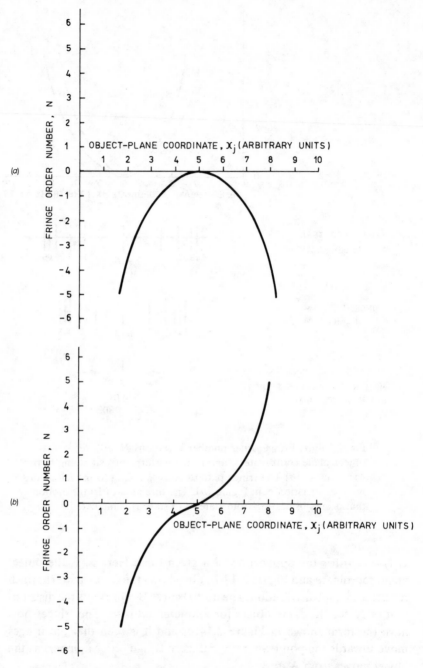

Fig. 2.15 (*a*), (*b*) and (*c*) Different fringe order number variations giving the same fringe patterns and motions as those shown in Figure 2.14 (*a*) and (*b*)

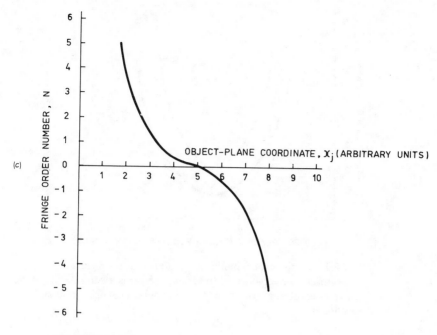

Fig. 2.15 (*cont.*)

by Figure 2.14(*b*) and (*c*). In each of these distributions, the magnitude of N varies in the same way. The modulus of N for a given point in an object undergoing a deformation of this form can be found by counting the number of fringes which pass that point as the load is applied.

Such a measurement does not indicate the sign of N. If, however, an additional load is applied giving a small increment δN whose sign is known, the sign of the gradient of N can be found. It is seen from Figure 2.14(*a*) that when $\partial N/\partial x_j$ and δN are positive the fringes move in the negative x_j direction, and when $\partial N/\partial x_j$ is negative and δN is positive, the fringes move in the positive x_j direction.

Thus, when δN is positive, the fringes due to the distribution in Figure 2.14(*a*) move towards the centre while those in Figure 2.15(*a*) move away from the centre. The distributions of Figure 2.15(*b*) and (*c*) give fringe motion to the left and to the right respectively. Thus these distributions can be distinguished from one another when the sign of δN is known. In practice a change in fringe order number can be obtained by applying gentle finger pressure at a point in the object surface. (Alternatively, one can determine the direction of displacement at a point using conventional techniques such as dial gauge deflection

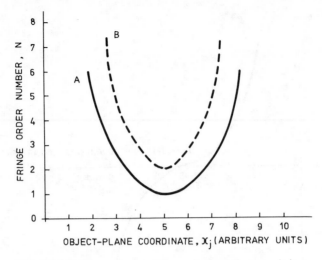

Fig. 2.16 Fringe order number variations that result in a symmetrical convergence of the fringes towards a central, non-static fringe at coordinate $x_k = 5$, as the distribution changes from state A to state B

readings for out-of-plane motion and strain gauges for in-plane displacement gradients.)

For the previous examples, the $N = 0$ position can be easily identified since the pattern appears static at that point. A fringe order distribution given by the continuous and dotted curves in Figure 2.16, gives rise to a fringe pattern which still converges to the centre but in this case the pattern there no longer appears static. The motion is difficult to describe, but can be visualized by imagining the fringes vanishing into a 'hole' located at the point towards which they converge.

We can conclude therefore that if the deformation of the object is such that all the fringes converge towards a single point or line which may be on or off the object surface, then the magnitude of N, the fringe order number, can be found for a given point and a given deforming load by counting the number of fringes passing that point as the load is applied. The sign of N may be found by observing the direction of fringe motion when a small additional load of known direction is applied. Many fringe patterns will be of this form, and can be readily identified as such by observing the overall fringe motion.

If, however, the distribution of N with x_k is such that its magnitude reaches a peak and falls off again, the fringe motion becomes more complex. In this case the fringes diverge from the position of the peak.

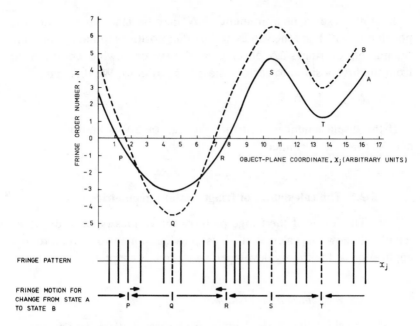

Fig. 2.17 Fringe patterns and motions associated with a general variation of N with respect to x_k

The distribution A shown in Figure 2.17 gives rise to a fringe pattern of the form shown below the graph. As the distribution changes from A to B, the direction of motion of the fringes is as indicated by the arrows. In the region around P, the fringes converge, and the point to which they converge moves to the right as the load increases. The fringes diverge from Q to both P and R while R moves to the left as the load increases. The fringes diverge from S to R and T, while the fringes at T will be observed to vanish down a 'hole'. In spite of its complexity, a displacement giving rise to a distribution of N having the form of Figure 2.17 can be evaluated if due account is taken of the various directions of fringe motion, and the sign of the gradient $\partial N / \partial x_j$ is determined.

It is clear from the discussion in Sections 2.4.1 and 2.4.2 that the procedure outlined above must be repeated for each of the three viewing directions when a full holographic fringe pattern analysis is required. It is important to note that fringe order numbers in live speckle pattern correlation interferometry (for example Sections 3.6.2 and 4.2) may also be determined using the same basic techniques: the procedure is somewhat simpler since the fringes can be made independently sensitive to out-of-plane and in-plane displacements.

In some cases a measurement of N may be made at one viewing position (say N_1) and the values at the other positions found by counting the number of fringes ΔN observed as the eye or viewing lens is moved from the first position to each of the other positions. We then have

$$N_{2,3} = N_1 \pm \Delta N_{2,3} \tag{2.45}$$

If the displacement is such that the sign of ΔN is known, then N_2 and N_3 can be determined.

2.5 The calculation of fringe pattern geometries

The forms of the fringe patterns for various types of displacement will now be calculated. This provides an introduction to the application of the theory developed so far in this chapter.

2.5.1 *Rigid body translations*

Consider first a rigid body displacement $(0, d_2, 0)$ i.e. one in which the displacement d_1 and d_3 are both zero so that only a displacement parallel to the x_2 axis exists. In accordance with the discussion in Section 2.3.3 the fringe pattern will localize at infinity. Values for the fringe spacings can be calculated from the following equations derived from the results in Table 2.1

$$\text{OV}_1 \text{ direction:} \qquad \Delta x_2 = \frac{\lambda R}{d_2}, \qquad \Delta x_3 = \infty \tag{2.46a}$$

$$\text{OV}_2 \text{ direction:} \qquad \Delta x_2 = \frac{-\lambda R}{d_2 \cos^2 \theta}, \qquad \Delta x_3 = \infty \tag{2.46b}$$

$$\text{OV}_3 \text{ direction:} \qquad \Delta x_2 = \frac{\lambda R}{d_2}, \qquad \Delta x_3 = \infty \tag{2.46c}$$

Thus, in each direction fringes parallel to x_3 axis are observed as shown in Figure 2.18. A displacement of 100 μm will give a fringe spacing of approximately 2.5 mm when $R = 500$ mm and $\lambda = 633$ nm. The same result is obtained for the Δx_3 components when the displacement is $(0, 0, d_3)$.

If the displacement consists of the components d_2, d_3 we find

$$\text{OV}_1 \text{ direction:} \qquad \Delta x_2 = \frac{\lambda R}{d_2}, \qquad \Delta x_3 = \frac{\lambda R}{d_3} \tag{2.47a}$$

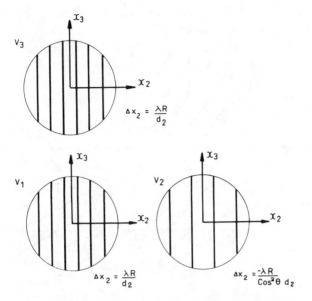

Fig. 2.18 Fringe patterns observed at V_1, V_2 and V_3 (see Figure 2.12) as a result of a displacement $(0, d_2, 0)$.

OV$_2$ direction: $\quad \Delta x_2 = \dfrac{-\lambda R}{d_2 \cos^2 \theta}, \quad \Delta x_3 = \dfrac{\lambda R}{d_3} \qquad$ (2.47b)

OV$_3$ direction: $\quad \Delta x_2 = \dfrac{\lambda R}{d_2}, \quad \Delta x_3 = \dfrac{-\lambda R}{d_3 \cos^2 \theta} \qquad$ (2.47c)

In this case diagonal fringes are obtained. Their orientation can be determined if we note that for the V_1 viewing position where $C_{12,2} = C_{13,3}$ the fringe order number variation will have maximum value in a direction parallel to the displacement vector, so that the fringes will be perpendicular to the displacement vector. The same sense of fringe orientation will be observed at V_2 and V_3 although the angle of inclination will change as a result of the change in the sensitivity to the d_2 and d_3 components respectively. This is illustrated in Figure 2.19 for the case where d_2 and d_3 are both positive and of the same magnitude.

Consider now a uniform translation d_1 perpendicular to the plane of the surface (d_2 and d_3 are assumed to be zero). In this case, the approximations made in Table 2.1 would suggest that no fringes are obtained in the OV$_1$ direction. When the exact expressions for C_{11} and $C_{11,2}$ are used it is found that circular fringes should be obtained. However, if the $N = 1$ fringe is to have a radius of 10 mm with $R = 500$ mm the

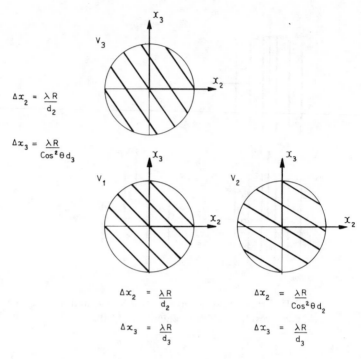

$$\Delta x_2 = \frac{\lambda R}{d_2}$$

$$\Delta x_3 = \frac{\lambda R}{\cos^2 \theta \, d_3}$$

$$\Delta x_2 = \frac{\lambda R}{d_2}$$

$$\Delta x_3 = \frac{\lambda R}{d_3}$$

$$\Delta x_2 = \frac{\lambda R}{\cos^2 \theta \, d_2}$$

$$\Delta x_3 = \frac{\lambda R}{d_3}$$

Fig. 2.19 $(0, d_2, d_3)$ displacement fringe patterns

displacement d_1 must be 25 mm. Thus, it is seen that this configuration is very insensitive to uniform translations in the d_1 direction.

The fringe spacings for observation in the OV_2 and OV_3 directions are given by:

$$\text{OV}_2 \text{ direction:} \qquad \Delta x_2 = \frac{\lambda X_{21}}{d_1 \tan \theta / (1 + \tan^2 \theta)^{\frac{3}{2}}}, \quad \Delta x_3 = \infty$$

$$(2.48a)$$

$$\text{OV}_3 \text{ direction:} \qquad \Delta x_2 = \infty: \qquad \Delta x_3 = \frac{\lambda X_{31}}{d_1 \tan \theta / (1 + \tan^2 \theta)^{\frac{3}{2}}}$$

$$(2.48b)$$

In this case a displacement of 100 μm observed at a viewing distance of 500 mm and an angle $\theta = 30°$ will give fringes of spacing approximately 4 mm.

A combination of the three displacements will obviously give fringe patterns combining these three effects. The normal view fringe pattern will be effectively sensitive to translations in the plane (d_2 and d_3) only.

2.5.2 An in-plane displacement gradient $(d_{2,3}, d_{3,2})$

Using the results in Table 2.1 we find that for observations at V_1

$$\Delta x_2 \to \infty \qquad (2.49a)$$

$$\Delta x_3 \to \infty \qquad (2.49b)$$

whilst for observations at V_2

$$\Delta x_2 \to \infty \qquad (2.50a)$$

$$\Delta x_3 = \frac{\lambda}{d_{2,3} \sin \theta} \qquad (2.50b)$$

and for observations at V_3

$$\Delta x_2 = \frac{\lambda}{d_{3,2} \sin \theta} \qquad (2.51a)$$

$$\Delta x_3 \to \infty \qquad (2.51b)$$

Equations 2.49 indicate that no fringes will be observed at V_1. This leads to the general result that when an object is viewed and illuminated in the same direction there is zero sensitivity to in-plane displacement gradients. Fringes lying parallel to the x_3- and x_2-axes will be observed at V_2 and V_3 respectively. It follows from the discussion in Section 2.3.1 that these fringes will localize in the plane of the object since only a displacement gradient is present. When $d_{2,3} = -d_{3,2}$ the displacement is equivalent to an in-plane rotation of the body. It can be seen that for $d_{2,3} = -d_{3,2} = 2 \times 10^{-4}$ radians and for $\theta = 30°$ we will have $\Delta x_2 = \Delta x_3 = 6.3$ mm when $\lambda = 633$ mm and the fringe patterns will have the form shown in Figure 2.20. Fringes of spacing of approximately 1 mm will result from an in-plane displacement gradient of about 1.2×10^{-3} radians.

2.5.3 An out-of-plane displacement gradient $d_{1,3}$

This is sometimes referred to as a tilt motion since the object rotates about a horizontal axis in the object plane. It is clear from Table 2.1 that fringes lying parallel to the x_2 axes will be observed at V_1, V_2 and V_3 where at V_1:

$$\Delta x_2 \to \infty \qquad (2.52a)$$

$$\Delta x_3 = \frac{\lambda}{2d_{1,3}} \qquad (2.52b)$$

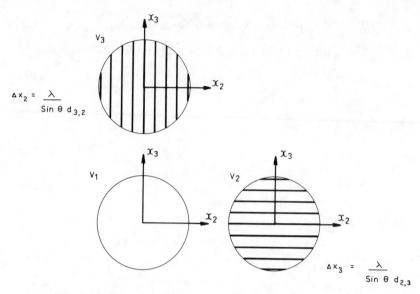

Fig. 2.20 $(d_{2,3}, d_{3,2})$ displacement fringe patterns

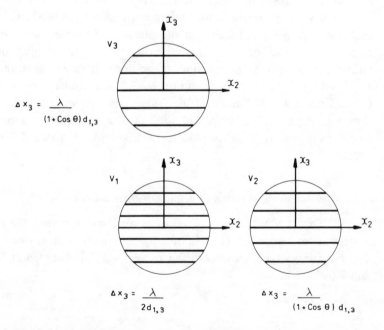

Fig. 2.21 $d_{1,3}$ displacement fringe patterns

at V_2

$$\Delta x_2 \to \infty \qquad (2.53a)$$

$$\Delta x_3 = \frac{\lambda}{(1 + \cos \theta)d_{1,3}} \qquad (2.53b)$$

and at V_3

$$\Delta x_2 \to \infty \qquad (2.54a)$$

$$\Delta x_3 = \frac{\lambda}{(1 + \cos \theta)d_{1,3}} \qquad (2.54b)$$

The fringe patterns, localized in the object plane, therefore have the form shown in Figure 2.21. By comparison with the example in 2.5.2 we find that $d_{1,3}$ needs to have a value of about 3.3×10^{-4} radians to produce fringes of 1 mm spacing when $\theta = 30°$. At this viewing angle the fringe pattern is therefore roughly three times more sensitive to out-of-plane than in-plane displacement gradients.

2.5.4 *A rigid body displacement* $(d_1, 0, 0)$ *superimposed upon an out-of-plane rotation* $d_{1,2}$

The three sets of equations for this combination of rigid body displacement and displacement gradient are as follows:

Viewing at V_1

$$\Delta x_2 = \frac{\lambda}{2d_{1,2}} \qquad (2.55a)$$

$$\Delta x_3 \to \infty \qquad (2.55b)$$

Viewing at V_2

$$\Delta x_2 = \frac{\lambda}{[d_1 \tan \theta / X_{21}(1 + \tan^2 \theta)^{\frac{3}{2}}] + d_{1,2}(1 + \cos \theta)} \qquad (2.56a)$$

$$\Delta x_3 \to \infty \qquad (2.56b)$$

Viewing at V_3

$$\Delta x_2 = \frac{\lambda}{(1 + \cos \theta)d_{1,2}} \qquad (2.57a)$$

$$\Delta x_3 = \frac{\lambda}{d_1 \tan \theta / X_{31}(1 + \tan^2 \theta)^{\frac{3}{2}}} \qquad (2.57b)$$

Since only two variables d_1 and $d_{1,2}$ are involved the components of the displacement can clearly be determined from the fringe patterns observed in any combination of two out of the three available directions. Let us consider the OV_2 direction in more detail. The fringes will be parallel to the x_3 axis and localize in the plane for which $\partial N_2/\partial x_k$ (equation 2.26) is a minimum. From an extension of the results in Table 2.1 to a point in a general plane we have

$$\frac{\partial N_2}{\partial x_1} = -\frac{d_1 \tan \theta}{X_1(1+\tan^2 \theta)^{\frac{3}{2}}} \qquad (2.58a)$$

$$\frac{\partial N_2}{\partial x_2} = \frac{d_1}{X_1}\frac{\tan \theta}{(1+\tan^2 \theta)^{\frac{3}{2}}} + (1+\cos \theta)d_{1,2} \qquad (2.58b)$$

$$\frac{\partial N_2}{\partial x_3} = 0 \qquad (2.58c)$$

where X_1 is the x_1 coordinate of the point in the plane.

Equation (2.58a) will be zero for $X_1 = \pm\infty$ i.e. the usual localization condition for rigid body translations will apply. However equation (2.58b) will be zero when

$$X_1 = -\frac{d_1}{d_{1,2}}\frac{\tan \theta}{(1+\tan^2 \theta)^{\frac{3}{2}}(1+\cos \theta)} \qquad (2.59)$$

The sign of X_1 above will depend on the signs of d_1 and $d_{1,2}$. It is clear therefore that the overall localization condition cannot be satisfied for a single value of X_1. In practice, fringes localizing either in front of the object (X_1, equation (2.59), positive) or behind the object (X_1, equation (2.59), negative) are observed. These fringes have reduced visibility due to the presence of the $\partial N_2/\partial x_1$ variation.

2.5.5 The in-plane strain of an object subject to a uniform axial load

Consider an object of the geometry shown in Figure 2.22 lying in the x_2x_3-plane. This is in the form of a 'thin' strip of material of uniform cross-section i.e. one in which the thickness is considerably less than the length, l, or width, w, and for which the ratio l/w is typically in the range 5 to 10. The resultant displacement is then described quite accurately by the equations

$$d_{2,2} = \varepsilon_{22} \qquad (2.60a)$$

$$d_{3,3} = \varepsilon_{33} = \nu\varepsilon_{22} \qquad (2.60b)$$

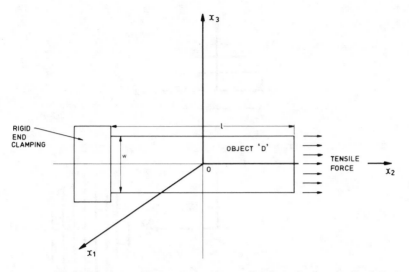

Fig. 2.22 An arrangement for the application of a longitudinal in-plane strain

$$d_{2,3} = d_{3,2} = 0 \qquad (2.60c)$$

where

$$d_{1,2}, d_{1,3} \ll d_{2,2}, d_{3,3} \qquad (2.60d)$$

ν is the Poisson's ratio of the material and ε_{22}, ε_{33} are the orthogonal linear strains in the plane of the specimen (11). This type of displacement will lead to the following forms of fringe patterns:

For observation at V_1

$$\Delta x_2 \to \infty \qquad (2.61a)$$

$$\Delta x_3 \to \infty \qquad (2.61b)$$

For observation at V_2

$$\Delta x_2 = \frac{\lambda}{d_{2,2} \sin \theta} \qquad (2.62a)$$

$$\Delta x_3 \to \infty \qquad (2.62b)$$

For observation at V_3

$$\Delta x_3 \to \infty \qquad (2.63a)$$

$$\Delta x_3 = \frac{\lambda}{d_{3,3} \sin \theta} \qquad (2.63b)$$

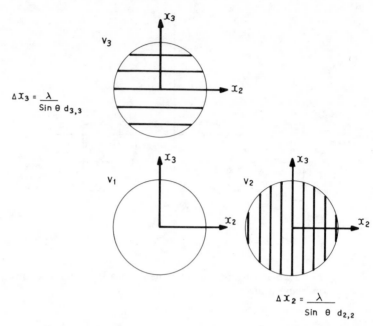

Fig. 2.23 Fringe patterns observed for an in-plane strain ($\varepsilon_{2,2}$, $\varepsilon_{3,3}$)

Since there is only a displacement gradient present the fringes will localize in the object plane (Section 2.3.3) and will appear as shown in Figure 2.23. Values of ε_{22} and ε_{33} can be determined from the fringe patterns viewed along the OV_2 and OV_3 directions respectively.

Equations (2.60) require that the loading force be applied in a direction precisely parallel to the x_2 axis and that there is no deformation of the jig holding the component. In practice these conditions are not usually fully satisfied and in addition to undergoing linear strain the specimen tends to bend and rotate. The holographic interferograms are sensitive to these components of displacement and they must all be determined independently from the analysis of the three patterns. This process can introduce substantial error and illustrates one of the basic limitations of holographic interferometry (Section 2.7.1).

2.5.6 *The deformation of a plane strip as a result of four-point bending*

Consider Figure 2.24. A rectangular strip of uniform cross-section is rigidly supported at equal distance from its centre by the supports S_1 and S_2 and is symmetrically loaded by applying equal forces at P_1 and

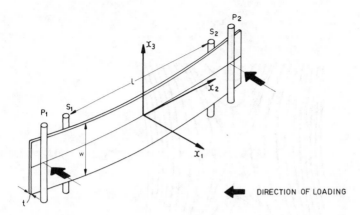

Fig. 2.24 An arrangement for the application of four-point bending

P_2. The application of load in this manner is known as four-point bending and results in the following components of surface displacement (12):

$$d_1 = -\frac{1}{2R_2}[x_2{}^2 + \nu(t^2 - x_3{}^2)] \tag{2.64a}$$

$$d_2 = \frac{M}{EI}x_2x_1 \tag{2.64b}$$

$$d_3 = -\frac{M\nu}{EI}x_3x_1 \tag{2.64c}$$

where

R_2 = the radius of curvature along the x_2-axis;
M = the applied bending moment;
ν = the Poisson's ratio of the material;
t = the thickness of the strip;
E = the Young's modulus of the material;
I = the moment of inertia of the cross-section of the strip.

When the interferogram corresponding to this deformation is viewed in the normal OV_1 direction it will be sensitive only to the components of out-of-plane strain as defined by $d_{1,2}$, $d_{1,3}$ displacement gradients. Under these conditions, we may regard it as mapping out contours of constant out-of-plane displacement at intervals of $\frac{1}{2}\lambda$. We have, from equation (2.39a), and Table 2.1,

$$N_1\lambda = 2d_1 \tag{2.65}$$

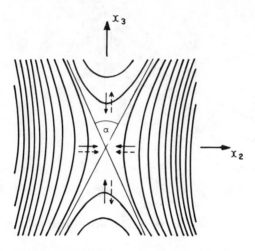

Fig 2.25 Fringes observed at V_1 due to the application of four-point bending shown in Figure 2.24

If we combine this result with equation $(2.64a)$ it can be seen that the fringes will consist of two sets of hyperbolae defined by the equation

$$x_2{}^2 - \nu x_3{}^2 = 0 \qquad (2.66)$$

This form of fringe pattern is shown in Figure 2.25 where the smallest angle, α, between the hyperbolas satisfies the relationship

$$\nu = \tan^2 \tfrac{1}{2}\alpha \qquad (2.67)$$

The interferometric observation of this type of deformation therefore provides a useful technique for the measurement of Poisson's ratio (13, 14). Fringe motion associated with four-point bending can be derived from the analysis in Section 2.4.4. When the loading is perfectly symmetric this is in the direction of the solid arrows shown in Figure 2.25, i.e. the fringes converge toward the static zero order fringe that exists along the asymptotes to the hyperbolas. Usually the surfaces undergo small rigid body rotations as well as anticlastic curvature due to small errors in the orientation of the load application system. For example, when the $d_{1,2}$ and $d_{1,3}$ components of the anticlastic displacement are respectively negative and positive and a negative tilt $d_{1,3}$ also occurs, the fringes move in the direction of the dashed arrows.

2.6 The measurement of dynamic displacements using holographic interferometry

In the previous sections it has been assumed that the original and displaced states are static (quasi-static displacements where the variation of position with respect to time is sufficiently slow for the resultant fringe motion to be resolved by the eye may also be included in this category). It will now be shown how the basic principles of holographic interferometry can be extended to the measurement of dynamic displacements. Three basic techniques may be used:

(a) time-averaged holographic interferometry (1);
(b) stroboscopic holographic interferometry (15);
(c) dual pulsed holographic interferometry (6).

(a) and (b) are suitable for the measurement of small amplitude periodic displacements and may be carried out using a continuous source laser. (c) may be used to observe considerably larger amplitude periodic displacements as well as non-periodic motions. These three methods will now be discussed in detail.

2.6.1 *Time-averaged holographic interferometry*

In time-averaged holographic interferometry the object is allowed to vibrate whilst the hologram is being exposed. Let us assume that a plane object is vibrating about a mean position with a periodic out-of-plane displacement $d_1(t) = a_0 \cos \omega t$ where a_0 is the amplitude of the vibration and $f = 2\pi/\omega =$ frequency. A holographic recording of this object is made at the normal viewing position, for example V_1, Figure 2.12. It can be seen that the light field scattered by the object at a time t is given by $U_s(t)$ where

$$U_s(t) = U_s \exp i[4\pi a_0 (\cos \omega t)/\lambda]$$

and

$$U_s = u_s \exp i\,\phi_s$$

(U_s is the complex amplitude of the light scattered from the stationary object.) If the complex amplitude of the holographic reference wavefront is now written as

$$U_r = u_r \exp i\,\phi_R$$

then the intensity in the hologram plane (Section 1.7.3) at a time t is given by $I(t)$ where

$$I(t) \propto U_r U_r^* + U_s(t) U_s^*(t) + U_r^* U_s(t) + U_r U_s^*(t)$$

$$= u_r^2 + u_s^2 + U_r^* U_s \exp \mathrm{i}\,[4\pi a_0(\cos \omega t)/\lambda]$$

$$+ U_r U_s^* \exp -\mathrm{i}\,[\pi a_0(\cos \omega t)/\lambda]$$

The hologram exposure time is made greater than the period of vibration $\tau = 2\pi/\omega$ so that the hologram effectively records a continuous distribution of frozen fringe holograms corresponding to the object at various points in its vibration cycle. Under these conditions the actual hologram intensity is given by the average $\langle I \rangle$ of $I(t)$ over the period where

$$\langle I \rangle = \frac{1}{\tau} \int_0^\tau I(t)\,\mathrm{d}t \propto u_r^2 + u_s^2 + \frac{U_r^* U_s}{\tau} \int_0^\tau \exp \mathrm{i}[4\pi a_0(\cos \omega t)/\lambda]\,\mathrm{d}t$$

$$+ \frac{U_r U_s^*}{\tau} \int_0^\tau \exp -\mathrm{i}[4\pi a_0(\cos \omega t)/\lambda]\,\mathrm{d}t \qquad (2.68)$$

The third term of the above equation is of primary importance to us since this governs the form of the virtual reconstructed image (see Section 1.7.3). It can be seen that

$$\int_0^\tau \exp \mathrm{i}[4\pi a_0(\cos \omega t)/\lambda]\,\mathrm{d}t$$

$$= \frac{1}{\omega} \int_0^{2\pi} \exp \mathrm{i}[4\pi a_0(\cos \omega t)/\lambda]\,\mathrm{d}(\omega t) \qquad (2.69)$$

By using the result (for example, reference 16) that

$$\frac{1}{2\pi} \int_0^{2\pi} \exp \mathrm{i}(x \cos \alpha)\,\mathrm{d}\alpha = J_0(x) \qquad (2.70)$$

where $J_0(x)$ is the zero order Bessel function, equation (2.69) becomes

$$\frac{1}{\omega} \int_0^{2\pi} \exp \mathrm{i}[4\pi a_0(\cos \omega t)/\lambda]\,\mathrm{d}(\omega t) = \frac{2\pi}{\omega} J_0(4\pi a_0/\lambda)$$

When the hologram is re-illuminated by the reference beam it therefore follows that the virtual reconstructed wavefront U_{rec} is given by

$$U_{rec} = U_r^2 U_s J_0(4\pi a_0/\lambda)$$

since $\omega\tau = 2\pi$ and the intensity of the reconstructed wavefront will be proportional to

$$U_{rec} U_{rec}^* = k J_0^2(4\pi a_0/\lambda) \qquad (2.71)$$

where k is a constant.

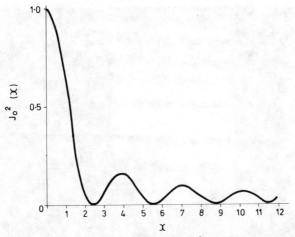

Fig 2.26 The first few cycles of the J_0^2 distribution which defines the intensity distribution of time-averaged holographic fringes for a sinusoidally vibrating surface

These results show that the intensity of the reconstructed image is modulated by a J_0^2 distribution. The first few cycles of this function are shown plotted in Figure 2.26.

It is clear from the graph that the brightest region of the reconstructed image ($x = 0$) will correspond to the nodal region in the vibration pattern i.e. where a_0 is zero. Subsequent maxima ($N = 1, 2, 3, \ldots$) will define contours of constant vibrational amplitude. Some values for the latter are listed in Table 2.2 from which it can be seen that the bright fringes

Table 2.2. *Vibration amplitude a_0 as a function of fringe order number N*

N	$a_0(\lambda)$
0	0
1.0	0.25
2.0	0.45
3.0	0.66
4.0	0.86
5.0	1.06
6.0	1.27
7.0	1.47
8.0	1.68
9.0	1.88

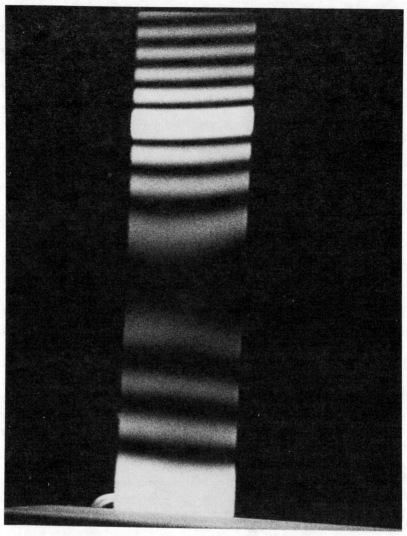

Fig. 2.27 Time averaged hologram of a strip vibrating in the modes shown in Figure 2.28. The strip lies in the x_2x_3-plane and the fringes define contours of constant out-of-plane vibration amplitude

will occur when a_0 is equal to approximately $\frac{1}{4}N\lambda$. A time-averaged interferogram of a metal strip rigidly clamped at one end vibrating in the mode drawn in Figure 2.28 is shown in Figure 2.27. The fall-off in fringe contrast characteristic of the J_0^2 intensity distribution is clearly visible. This interferogram was observed in the V_1 viewing position. When observation is carried out at an angle to the surface normal (for

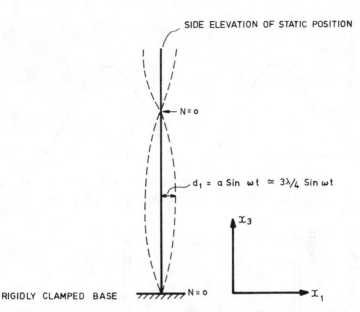

SIDE ELEVATION OF STATIC POSITION

N = o

$d_1 = a \sin \omega t \simeq 3\lambda/4 \sin \omega t$

x_3

RIGIDLY CLAMPED BASE   N = o x_1

Fig. 2.28 Vibration mode corresponding to the out-of-plane vibration amplitude and time-averaged interferogram shown in Figure 2.27. The nodal regions (regions of maximum intensity) in Figure 2.27 enable the mode shape to be identified

example V_2, Figure 2.12) the distribution would be $J_o^2[4\pi a_o(1 + \cos\theta)/\lambda]$ due to the change in the C_i coefficient.

2.6.2 *Stroboscopic holographic interferometry*

Time-averaged holographic interferometry suffers from two disadvantages. Firstly it is a frozen fringe technique and cannot therefore be used to detect resonant modes in real time,[*] and, more seriously, the phase of the vibration components is lost due to the averaging of the displacements over a vibration cycle. Both these limitations are eliminated if stroboscopic holography is used.

In this technique, a hologram is made of the object while it is stationary and the hologram is relocated to give a 'fluffed out' fringe pattern as described in Section 2.1.

---

[*] The speckle pattern interferometric technique described in Sections 3.7.4 and 4.7.1 enables time-averaged fringe patterns to be observed in real time. Sometimes these methods are used to locate the required resonant mode which is then recorded holographically. (See for example Section 6.6.1.) The advantage of the holographic recording is that it enables better contrast fringes to be obtained.

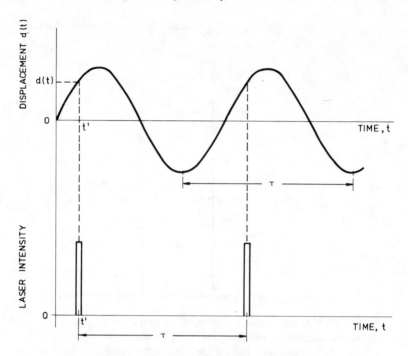

Fig. 2.29 The form of the intensity modulation for the real-time stroboscopic observation of the sinusoidal displacement $d(t)$

The object is made to vibrate and the intensity of the object beam is modulated so that the object is illuminated by a series of pulses which occur at the same frequency as that at which the object is vibrating and whose duration is much less than the period of the vibration (Figure 2.29). If the object motion at a point r is described by

$$a(r) = a_0(r) \cos [\omega t + \phi(r)] \tag{2.72}$$

and the pulses occur at times

$$t = t', t' + 2\pi/\omega, t + 4\pi/\omega, \ldots$$

then the displacement at the point r each time it is illuminated is given by

$$d_{st} = a(r) \cos [\omega t_0 + \phi(r)] \tag{2.73}$$

The hologram reconstructs the wavefront scattered from the object in its original stationary position so that the interference fringes will represent contours of constant d_{st}. (The persistence of the eye or a television camera will give the appearance of continual illumination.)

Values of $\phi(r)$ at a given point in the pattern can be determined by measuring the value of t' which gives a minimum in the intensity at that point.

2.6.3 *Dual pulsed holographic interferometry*

When making a hologram it is necessary that all the components are stable to considerably better than a wavelength during the holographic exposure time which for continuous source lasers is typically of the order of several seconds. The output of a pulsed laser (7) may compress an equivalent amount of energy into a pulse of duration $\simeq 10^{-8}$ s and a dual pulsed laser can produce two such pulses with a separation of $\simeq 10^{-5}$ s, so that the use of such a laser considerably relaxes the stability requirements of holographic interferometry. Pulse energies in excess of 1 joule may be delivered by a ruby laser at a wavelength of 654 nm – see Table 6.2.

A hologram of an object made using a dual pulsed laser will give a frozen fringe pattern which represents the displacement undergone by the object in the time interval between the two pulses. Such a hologram can be reconstructed using a continuous output laser. A HeNe laser (Table 6.2) with a wavelength of 633 nm is normally used for this purpose. The small wavelength difference between the HeNe and the ruby laser light used to record the hologram means that only a small image distortion is introduced. For the purposes of fringe observation and interpretation this may be ignored. The interferogram shown in Figure 2.30 was recorded by Gates, Hall and Ross (6) in this way. It shows the deflection of a crash helmet which was struck by a hammer whilst being worn by a live model. The helmet was illuminated after the impact by two 25 ns pulses separated in time by 25 μs delivered by a dual pulsed ruby laser. It is interesting to note that there are no fringes on the face of the model. This indicates the insensitivity of the recording method to random, low velocity motions.

The technique can also be used to measure large amplitude periodic displacements. If, for example, the period of oscillation is 10^{-3} s a pulse separation of 10 μs will enable 0.01 of the overall amplitude to be measured in a single recording. This will extend the range of vibration measurement to amplitudes approaching $10^{3}\lambda$.

Holographic interferometers for use with a pulsed laser are similar in layout to that of a conventional CW system but concave lenses must be used to expand the laser beams, since when high energy pulsed beams are brought to a focus by a convex lens an explosion may occur at the

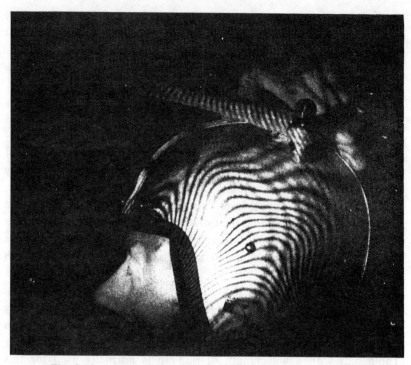

Fig. 2.30 A dual pulsed holographic interferogram showing the
impact deflection of a crash helmet worn by a live model. The pulse
separation for this recording was 25 μs. (Reproduced by kind
permission of Dr John Gates, see Gates, Hall and Ross (6))

point of focus due to the ionization of the air. This will spoil the laser
beam shape and the hologram will not be recorded.

Some practical aspects of pulsed system design and alignment are
also given in Section 6.6.1.

2.7 Holographic interferometry at increased sensitivity

The intensity in the viewing plane of a holographic interference
pattern can be written as

$$I = I_\text{o}(1 + \cos \phi)$$

see equation (2.19), where $\phi = 2\pi/\lambda\,(n_\text{o} - n_\text{s}) \cdot d$ and d represents the
change in position of the object which is to be determined. The variation in
ϕ is generally found by locating fringe minima and/or maxima. The
accuracy to which these may be located is limited by the optical noise

in the interferogram and by the overall variation in the intensity, I_o, of the object beam. Maximum resolution is typically one-tenth of a fringe which limits the accuracy with which d can be determined.

2.7.1 *Heterodyne holographic interferometry*

Heterodyne holographic interferometry (HHI) is a technique developed by Dändliker, Inleicher and Hottier (17) which enables the value of ϕ to be found to considerably higher accuracy. In this technique, the two object wavefronts corresponding to the object in its two positions have different frequencies. When two waves of different frequencies are added together, it is possible under certain conditions to measure their relative phase very accurately. This is seen as follows.

When two waves of different frequencies f_1, f_2 having amplitudes and phases u_1, ϕ_1 and u_2, ϕ_2 respectively are added together, the resultant intensity is given from equation (1.23) as

$$I(t) = u_1^2 + u_2^2 + 2u_1u_2 \cos\left[2\pi(f_1 - f_2) + (\phi_1 - \phi_2)\right]$$

In holographic interferometry, u_1 and u_2 are generally made equal, so that

$$I(t) = 2u_1^2\{1 + \cos\left[2\pi(f_1 - f_2)t + (\phi_1 - \phi_2)\right]\}$$
$$= 2u_1^2\{1 + \cos\left[2\pi Ft + \Phi\right]\} \tag{2.74}$$

where

$$F = f_1 - f_2$$

and

$$\Phi = \phi_1 - \phi_2$$

If the light is detected by a detector which can resolve the frequency F, the output signal will therefore have a mean value proportional to $2u_1^2$, and will be sinusoidally modulated at frequency F with amplitude $2u_1^2$. The phase of this signal varies as the value of Φ varies. When the phase Φ is compared electronically with that of a reference signal at frequency F, the variation in Φ and hence d across the field of view can be measured. Such a measurement can be made to an accuracy of typically 10^{-2} radians or 3×10^{-3} of a cycle and hence enables very small changes in d to be resolved. Moreover the measurement is independent of the value of u_1 which affects only the amplitude of the signal.

Heterodyning may be used in both live and double exposure holography. In live HHI, the 'live' object beam and the reconstructing reference beam have different frequencies, whilst in double exposure HHI, a single frequency is used to make both exposures but two separate reference beams are employed. At the reconstruction stage these two reference beams are set at different frequencies. In both methods two detectors are required which are used in one of two ways. For the first of these, one of the detectors is tracked across the fringe pattern whilst the other is held static and hence generates the fixed reference frequency. The variation in Φ across the fringe field is thereby obtained. Alternatively the two detectors may be maintained at a fixed distance with respect to one another and both tracked across the fringe field. This enables a differential measurement to be made and hence the variation in the gradient of Φ is found. In practice the two frequencies are obtained by splitting a single continuous laser beam and passing the two beams through separate opto-acoustical modulators. Decorrelation of the speckle due to the displacement (see for example, Appendix F.2) will give rise to a degree of phase measurement error. However, Dandliker (17) has shown that these errors can be reduced if the measurement of the intensity $I(t)$ is made over many speckles.

HHI has been used in these ways to measure very small displacements and to make very high resolution shape measurement. The phase of a double-exposure holographic interferogram made using a dual pulsed laser may be measured when a HeNe laser is used to reconstruct the interferogram. It cannot be employed, however, in time-averaged holographic interferometry since the phase is averaged over many cycles during the recording time.

Another way in which the sensitivity of holographic interferometry may be increased is by modulating the phase of the reference beam (18). The principle of this technique is the same as that explained in Section 4.7.2 where the same basic method has been used in conjunction with electronic speckle pattern interferometry.

2.8 A critical assessment of holographic interferometry

It should be clear from the discussion in this chapter that although holographic interferometry is in principle a relatively simple technique, the successful application of the method to quantitative measurement can be quite difficult. Some of the main problems are outlined in Section 2.8.1. There are, however, a large range of problems that can be investigated for which qualitative fringe interpretation is quite adequate. These are summarized in Section 2.8.2. Finally the main

factors that limit the range of application of holographic interferometry
are considered (Section 2.8.3).

2.8.1 *General comments on quantitative fringe pattern analysis*

Quantitative measurements in holographic interferometry are
concerned usually with the measurement of surface deformations. Rigid
body motions should therefore be minimized. This can be done by careful
design of the deformation jig and generally requires that a region of the
object be supported rigidly. The intrinsic sensitivity of the fringe pattern
to rigid body translation (for example, Section 2.4.3) can also be reduced
by making the viewing distance as large as practically possible. Under
these conditions the effects of rigid body translations can generally be
neglected and the interpretation of the fringe pattern in terms of displacement gradients becomes less complex (see also Section 2.4.3). Note that
the presence of any significant rigid body movement can usually be
detected since it will tend to make the fringes locate in a plane away
from that of the object (Section 2.5).

We have seen that for all but the normal V_1 viewing position
(Figure 2.12) the fringe patterns are sensitive to both out-of-plane
and in-plane displacement gradients. All the necessary displacement ·
information is contained in the three interferograms observed at V_1, V_2
and V_3. One is therefore left with the problem of separating the
out-of-plane displacements as observed at V_1 from the combinations
of in-plane and out-of-plane displacements observed at V_2 and V_3.
This can introduce substantial experimental errors. (The fact that
for practical viewing angles θ, the V_2 and V_3 interferograms are
typically two to three times more sensitive to out-of-plane than
in-plane displacements (Section 2.5) can be a further source of
difficulty when the in-plane displacements are small with respect to the
out-of-plane.) For these reasons accurate quantitative results
are most readily obtained for the V_1 viewing position. Direct
observations of in-plane displacement gradients are probably best
made using one of the speckle pattern interferometric techniques discussed in the next two chapters, for example Sections 3.2, 3.7.2 and 4.5.1.

In the case of dynamic displacement measurements time-averaged
interferograms may be readily interpreted for out-of-plane sinusoidal
vibrations. Nodal zones are well defined and the fringes simply define
contours of constant amplitude of vibration. A considerable number
of components of physical interest will be found to vibrate in this way
when mechanically excited by a sinusoidal wave and consequently the

technique provides a particularly powerful tool for experimental vibration analysis. The main disadvantages are the loss of phase information and the fact that resonant modes cannot be detected in real time.

The analysis of the fringe pattern becomes difficult however when either the vibration is not sinusoidal, or when it has both in- and out-of-plane components. When the first occurs, the envelope of the fringe visibility does not have the J_0^2 distribution of equation (2.71) (41), while the presence of in- and out-of-plane components gives rise to the problem of separating these components in the fringe pattern analysis – see Section 2.4.

If double-exposure interferograms (including dual pulsed fringe patterns) are to be interpreted fully, additional information concerning for example regions of zero or minimum displacement and displacement direction is required. This can sometimes be deduced from an analysis of the mechanical constraints present in the system. Such a procedure clearly introduces an additional area of uncertainty into the analysis, but it is nonetheless essential in the case of dual pulsed interferograms where it is not possible to observe the displacement in real time. When live fringe observation is possible, as in quasi-static displacement observation and stroboscopic holography, the required information can generally be derived directly from the movement of the fringes. However, successful real-time holography requires precise plate relocation (Section 6.5.2) and careful plate processing involving pre-exposure emulsion expansion to eliminate error fringes (Section 6.4.2). Such a procedure is quite time consuming, and whereas the technique may provide an invaluable research tool it is unsuitable for routine measurement. (The problem may, however, be overcome by the use of a thermoplastic recording material, see for example, Section 5.2.)

In the theory of the last sections we have assumed that the illumination wavefront is plane in order to avoid unnecessary complications, for it can be seen that if the wavefront is curved we must take into account the variation in the angle of illumination over the object. This results in a variation in the sensitivity to displacement gradients across the object plane and an increased sensitivity to rigid body translations. For these reasons it is preferable to use collimated illumination or an illumination wavefront of as large a radius of curvature as is practical.

2.8.2 *General comments on qualitative fringe pattern analysis*

It should not be concluded from this chapter that quantitative fringe pattern interpretation is always necessary, or even desirable. In that the technique is capable of being used as a precise measurement tool it is of course essential that we should be able to analyse the fringe patterns where required. There exist, however, a wide range of problems in which the qualitative appearance of the fringe pattern enables the phenomena under investigation to be visualized. For example such methods are often used in the following disciplines:

> non-destructive testing,
> engineering component design analysis,
> vibration mode detection.

Typical applications in the above areas are discussed in Chapter 7. The reason they have not been considered in this chapter is that there are no general solutions as in quantitative fringe pattern interpretation. Each investigation is governed by the specific nature of the problem and often requires a partially empirical approach. (This will become clear to the reader when the relevant sections of Chapter 7 are studied.)

The technique in which holographic interferometry is used to observe refractive index (and hence density) variations through a volume of fluid can be readily used as a qualitative tool for the detection of, for example, regions of maximum thermal gradient or fluid flow. Whether or not such patterns can be analysed quantitatively depends upon the complexity of the boundary conditions and the accuracy with which they can be modelled mathematically. A method of analysis for the fringe field corresponding to an axisymmetric refraction index variation is outlined in Section 7.7.

2.8.3 *Limiting factors in holographic interferometry*

Any object which scatters or reflects light can in principle be used to make a hologram which will reconstruct a wavefront equivalent to the original object wavefront. To make such a hologram in practice, the laser power must be sufficient to ensure that the interference fringes formed on the holographic recording medium between the object and reference beams are stationary during the recording of the hologram – or at least that any movement of these fringes is very much less than the fringe spacing (Section 6.4.1). The recording time is determined by the intensity of the object and reference beams. The amount of light scattered by the object onto the recording plane is a function of its size, shape and surface finish whereas the stability of the fringes depends on the rigidity of the object, and of the components in the optical system. (These factors are discussed in detail in Chapter 6.) The maximum area which can be viewed with a single hologram is governed basically by the amount of laser power available. Very small areas can also be observed but viewing of such areas can be difficult. Another limiting factor in making a hologram is the coherence length of the laser. This must be such that the path difference between the object and reference beams must be less than the coherence length. It may be necessary to use a single mode laser when making a hologram of objects whose depth varies considerably, or those of large area.

Whether or not interference fringes can be obtained between either the live and reconstructed wavefronts in live HI or between the two reconstructed wavefronts in double exposure HI depends on the magnitude and direction of the object displacement and on the size of the viewing lens aperture. It is shown in Appendix F.1 that when the ratio of the fringe spacing to the mean speckle size approaches unity, the visibility of the fringes decreases, and no fringes are observed when this ratio is unity or less. Thus, the viewing lens aperture should be as large as possible. When the object is not flat, the lens aperture size may be limited by the depth of focus required to view the whole surface. For a given numerical aperture, there will be a lower limit to the spacing of the HI fringes which can be observed and hence an upper limit to the displacement which can be detected.

The presence of rigid body translations causes the holographic fringes to be localized away from the object surface and it is then more difficult to relate the fringe positions to object coordinates. This is particularly inconvenient in non-destructive testing applications and is another reason why rigid body movements should be avoided as far as possible in making holographic interferograms. This becomes more difficult when small areas are being viewed, since the displacement gradient required to produce a given number of fringes increases as the object area decreases: the deforming force must then be greater and rigid body translations are more likely to arise. This is the major factor limiting application of this type in which the problems involved become mechanical rather than optical. Fringes over an area of about 500 μm due to the local depression of the surface have been observed (19).

Holographic systems using pulsed lasers which are capable of inspecting objects of up to 6 m width have been reported. An interferogram of a microwave antenna 3 m in diameter obtained by Wuerker is shown in reference (20), Figure 8.14.

2.9 Suggestions for further reading

In addition to the publications already referenced in this chapter it is suggested that the reader who wishes to increase his knowledge of the subject may find the following reading beneficial. For convenience the references have been grouped into specific categories.

(1) *Fringe pattern interpretation methods*: References (21)–(34). (Here the review article by Briers (34) is a useful starting point. References (25), (26) and (27) which describe respectively the Fourier viewing of holographic fringes and the holodiagram represent interesting departures from conventional fringe analysis methods.)

(2) *The extension of dynamic measurement techniques*: References (35)–(40).

(3) *Textbooks*: The review text edited by Erf (20) together with the Conference Proceedings (8, 42, 43) show how the state of the art has developed over the last 15 years. The textbook by Vest (44) has helped to consolidate a lot of this work.

Further references relating to applications of holographic interferometry are given in Chapter 7.

3

Speckle pattern interferometry

3.1 Introduction: a comparative summary of techniques

Two main techniques are grouped within the general classification of speckle pattern interferometry. These are:

(i) speckle pattern correlation interferometry; and
(ii) speckle pattern photography.

In both of these a fringe pattern is derived from an optically rough surface observed in its original and displaced positions. Depending upon the method of recording and fringe observation these fringe spacings can be made sensitive to the local displacements, displacement gradients (Sections 3.2 and 3.6) or the first derivative of the displacement gradient (Sections 3.3 and 3.7.2). As will become apparent, the directional and magnitude sensitivity of these fringes can also be varied over a substantially larger range than those in holographic interferometry. Furthermore the recording medium need not have such a high spatial resolution (for example Section 3.2.1). These factors combine to make speckle pattern interferometry a more flexible technique for displacement measurement than holographic interferometry despite the fact that fringe definition is usually poorer.

   The first of these techniques, speckle pattern correlation interferometry, was described initially by Leendertz (1) and indeed it was the need to overcome some of the inherent problems of holographic interferometry (for example, Section 2.8.1) that stimulated the early work. A general interest in the properties of speckle patterns (Section 1.8) together with the work of Groh (2), (483–94) influenced the initial experiments. Groh had used the relocated negative of an image-plane speckle pattern as a shadow mask as a means of detecting fatigue cracks. At about the same time Maron (3) independently performed a similar experiment with the introduction of an off-axis reference beam. In both cases the overall change in the correlation of the live and recorded speckle patterns was recorded during the development of a fatigue crack. Leendertz showed that the individual intensities of image-plane speckle could be made to vary cyclically for a given direction of object motion if they interfered with a reference beam of specific geometry. The analysis in Section 3.3 shows that the averaging of this effect over many speckles leads to the formation of fringe patterns of wavelength displacement sensitivity. Such a pattern is obtained by correlating the image-plane intensity distribution of the surface in its displaced and undisplaced positions. When this is done the directional sensitivity of the fringes is a function of the reference-beam geometry, which may be arranged to give out-of-plane or in-plane displacement sensitivity or out-of-plane displacement gradient sensitivity. An appreciation of the distinction between the formation of fringes by intensity correlation as described here and by wavefront interference (for example, Section 1.3.4) is basic to the understanding of the method.

In early experiments correlation fringes were observed by the super-position of a negative of the undisplaced speckle pattern upon a positive of the displaced-state speckle pattern. (The correlation fringes are thereby obtained by a process of intensity multiplication which is the same in principle as the live fringe method discussed in Section 3.6.2.) This is a somewhat cumbersome process and the results of Burch and Torkaski (4) combined with the general principle of Fourier filtering (Section 1.9) led to a single-plate double-exposure correlation technique (5, 6).

Part of these double-exposure investigations showed that displacement sensitive fringe patterns could be obtained from photographically recorded speckle patterns in the absence of a reference beam. This was an important discovery and now forms the basis of a wide range of displacement measurement techniques classified under (ii). In these methods, the object is illuminated by a single beam. A viewing lens collects some of the light scattered from the surface, and a photographic recording is made on a single plate of the light scattered from the original and displaced object positions. The recording plane may be at the plane in which the object is imaged, or may be at the focal or some intermediate plane (defocussed speckle). It may be shown (Sections 3.2 and 3.3) that image-plane speckle patterns generate fringe fields sensitive to in-plane displacements (5, 6, 7) whereas defocussed speckle patterns are pre-dominantly sensitive to out-of-plane displacement gradients. The first detailed analysis of the use of defocussed speckle patterns was carried out by Tiziani (8) and Gregory (9). Defocussed speckle effects are also discussed by Archbold and Ennos (7).

The magnitude sensitivity of the fringes using speckle photographic techniques can be varied between 0.1 μm and 100 μm; this will be shown in Section 3.5.2.

Two major factors which distinguish speckle pattern photographic (ii) from speckle correlation techniques (i) are the difference in displacement magnitude sensitivity and the absence of a reference beam at the recording stage. For brevity speckle pattern correlation (i) and speckle pattern photography (ii) will be identified by the initials SPC and SPP.

In Section 3.2 the basic principle of double-exposure SPP is discussed and in Sections 3.3 and 3.4 it is shown how the techniques may be extended to the measurement of out-of-plane displacement gradients (for example, surface tilt) and dynamic displacements. Speckle pattern correlation interferometry is dealt with in a similar way in Sections 3.6 and 3.7. The chapter also includes discussions of the limitations of the two techniques (Sections 3.5 and 3.8). Practical methods are outlined where it is felt that this helps to clarify the theoretical description.

3.2 Image-plane recording speckle pattern photography

In SPP a viewing lens collects light scattered from an object in an undisplaced and a displaced position, and a recording is made of the two scattered fields on a single photographic plate. The light in the recording plane scattered from the undisplaced and displaced objects will consist of two speckle patterns. These two speckle patterns can be assumed to be identical (the conditions for which this assumption is valid are discussed in Appendix F), except that one is displaced with respect to the other by an amount which depends on the size of the displacement of the object and by the position of the recording plane. Thus if the relative displacement of the two speckle patterns can be measured, the displacement of the object can be found.

It will be shown in Section 3.2.1 that the speckle displacement and hence the object displacement may be determined from the Fourier transform of the doubly exposed photograph and that this can be viewed by using one of the optical methods of mapping out the transform in a Fourier plane outlined in 1.9. These are as follows:

> (1) The plate is illuminated by a plane wavefront and the diffrac-
> ted light viewed in the focal plane of a lens.

(2) The plane is illuminated by a converging wavefront and the diffracted light viewed in a plane containing the point of convergence of the illuminating wavefront.
(3) The plate is illuminated by a plane wavefront and the diffracted light viewed in a plane which is sufficiently far away from the plate to approximate to a Fourier plane.

In Section 3.2.2, it is shown how point-by-point measurements of the displacement can be made, and in Section 3.2.3 a method is described which gives an image of the object upon which a fringe pattern defining the object displacement is superimposed.

Throughout Section 3.2, the focussed speckle case is described; i.e. the position of the recording plane is such that the object is in focus in that plane. The analysis in Section 3.3 shows that when the recording plane is such that the object is not in focus, the speckle motion is predominantly sensitive to surface tilt. Again, the basic Fourier viewing methods discussed in Sections 3.2.2 and 3.2.3 can be used to measure the magnitude and direction of the speckle motion and hence to measure the tilt.

3.2.1 *The basic Fourier viewing method for fringe observation*

Consider the arrangement shown in Figure 3.1. A plane optically rough surface D lying in the x_2x_3-plane is viewed normally in the x_1 direction by a lens L of focal length f and aperture diameter a. The surface is illuminated by a divergent wavefront, U_0, and a photograph of the image-plane speckle pattern is recorded in the plane I.

If the photographic plate is exposed once only, the intensity distribution of the speckle pattern can be represented as the sum of a set of

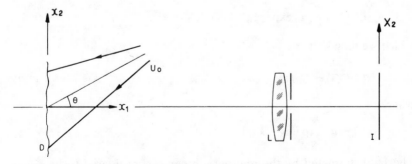

Fig. 3.1 The basic arrangement for the recording of an image plane speckle pattern in double-exposure SPP

sinusoidal gratings whose spatialfrequencies vary between zero and an upper limit, which is determined by the viewing lens aperture and the lens-to-recording plane distance (see Section 1.8.2). When the developed plate is illuminated by a plane wavefront and the diffracted light is observed in a Fourier plane, it will form a circle whose diameter is determined by the maximum spatial frequency. This is known as the diffraction halo; an example is shown in Figure 3.2(a).

When the plate is exposed first to light scattered from the undisplaced object, and then to light scattered from the object after it has been displaced by an amount d_2 in the x_2-direction, two identical speckle patterns are obtained which are displaced with respect to one another by an amount Q_2. In this case $Q_2 = md_2$ where m is the magnification of the viewing system, (Q_2 is also parallel to the x_2-axis). We see that a component grating in the first speckle pattern of spatial frequency f_{x_2} corresponding to a spacing $S = 1/f_{x_2}$, will be in anti-phase with the grating of the same spatial frequency in the displaced pattern when

$$Q_2 = (n + \tfrac{1}{2})S, \qquad n = 0, 1, 2, 3 \cdots \tag{3.1}$$

Thus, when the diffracted light is observed in the Fourier plane, no light will be seen at positions corresponding to spatial frequencies satisfying equation (3.1). This result is obtained quantitatively in the following analysis:

The intensity at a point $P(X_2)$ in a singly exposed speckle pattern in the X_2X_3-plane is given by $I(X_2)$ where, for simplicity, we consider only variations in the X_2-direction. When such a photograph is developed and illuminated by a normally incident plane wavefront of amplitude $u_0 \exp i(\phi_0 + kx_1)$ the amplitude of the transmitted light in the plane of the plate, $U_T(X_2)$, is

$$U_T(X_2) \propto [u_0 \exp i(\phi_0 + kx_1)]I(X_2)$$

which we may write as

$$U_T(X_2) = U_0 I(X_2) \tag{3.2}$$

where

$$U_0 = u_0 \exp i(\phi_0 + kx_1)$$

The light diffracted by the photograph can be considered to be made up of a set of plane waves (see Section 1.4.1) and the amplitude $U_{SE}(\sin \alpha)$ of the light component diffracted at an angle α with respect

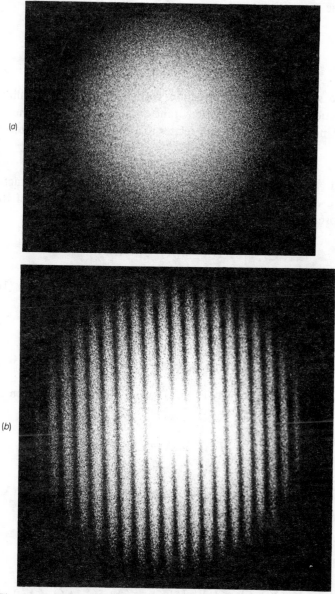

Fig. 3.2 (*a*) The diffraction halo obtained by illuminating a singly exposed image-plane speckle pattern recording with an unexpanded laser beam. (*b*) Diffraction halo fringes obtained by illuminating a double-exposure image-plane speckle pattern with an unexpanded laser beam. The object was shifted a distance $d_2 = 200$ μm in its plane between exposures. A general geometry for diffraction halo fringe formation from a double-exposure image-plane negative is shown in Figure 3.3

to the x_1-axis is given from equation (1.93) as

$$U_{SE}(\sin \alpha) = U_0 \mathscr{F}[IX_2]$$

$$= U_0 \int I(X_2) \exp [-2\pi i X_2(\sin \alpha)/\lambda] \, dX_2 \qquad (3.3)$$

since $\sin \alpha$ equals the direction cosine of a wave travelling in this direction. $\mathscr{F}$ is the Fourier transform operator, and the subscript 'SE' indicates that the result relates to a single exposure result. We have

$$I_{SE}(\sin \alpha) \propto |\mathscr{F}(U_T)|^2$$

where $I_{SE}(\sin \alpha)$ represents the intensity of the light diffracted at an angle α by a singly exposed photograph. The intensity of the singly exposed speckle pattern may be written in the form

$$I_{orig}(X_2) = \int_{-\infty}^{+\infty} I(X_2')\delta(X_2 - X_2') \, dX_2' \qquad (3.4)$$

In the above equation $\delta(X_2 - X_2')$ is a Dirac Delta function (10) which is defined by the conditions:

$$\int_{-\infty}^{\infty} \delta(X - X') \, dX' = 1$$

$$\delta(X - X') = 0 \quad \text{when } X \neq X'$$

We see that the intensity at $P(X_2)$ when the object has been displaced by a distance d_2 is given by

$$I_{disp}(X_2) = \int_{-\infty}^{\infty} I(X_2')[\delta(X_2 + Q_2 - X_2')] \, dX_2' \qquad (3.5)$$

where $Q_2 = md_2$.

The intensity of the doubly exposed speckle pattern at $P(X_2)$ can be written in the form $I_{DE}(X_2)$ where

$$I_{DE}(X_2) = I_{orig}(X_2) + I_{disp}(X_2) \qquad (3.6)$$

and hence

$$I_{DE}(X_2) = \int_{-\infty}^{+\infty} I(X_2')[\delta(X_2 - X_2') + \delta(X_2 + Q_2 - X_2')] \, dX_2'$$

$$(3.7)$$

The light diffracted at an angle α now has an amplitude $U_{DE}(\sin \alpha)$ where

$$U_{DE}(\sin \alpha) = U_0 \mathscr{F}[I_{DE}(X_2)] \qquad (3.8)$$

Equation (3.7) represents the convolution of the functions $I(X_2)$ and

$[\delta(X_2) + \delta(X_2 + Q_2)]$, and from the Convolution Theorem (see Appendix C, equations C.10 and C.11) we have

$$U_{DE}(\sin \alpha) = U_0 \mathcal{F}[I(X_2)] \cdot \mathcal{F}[\delta(X_2) + \delta(X_2 + Q_2)]$$

It can be shown (see for example, references 11, 12) that

$$\mathcal{F}[\delta(X_2) + \delta(X_2 + Q_2)] = \cos\left(\frac{\pi}{\lambda} Q_2 \sin \alpha\right) \exp\left(-i\frac{\pi}{\lambda} Q_2 \sin \alpha\right)$$

so that amplitude of the light diffrac: :d at an angle α is

$$U_{DE}(\sin \alpha) = U_{SE}(\sin \alpha) \cos\left(\frac{\pi}{\lambda} Q_2 \sin \alpha\right)$$

The intensity of the light in the Fourier plane is therefore given by

$$I_{DE}(\sin \alpha) = I_{SE}(\sin \alpha) \cos^2\left(\frac{\pi}{\lambda} Q_2 \sin \alpha\right) \tag{3.9}$$

Thus, the diffraction halo will contain fringes with minima occurring when

$$\sin \alpha = (2n + 1)\lambda / Q_2, \qquad n = 0, 1, 2, 3, \ldots \tag{3.10}$$

The spacing of these fringes can be seen to be equivalent to those that are obtained with two slits (see Section 1.6.3) and for this reason are often referred to as Young's fringes. If α is small, their angular separation is given by

$$\alpha = \lambda / Q_2 \tag{3.11}$$

(Note also that the fringe spacing in the halo will be $l\alpha$ where l is the distance between the plate and the Fourier plane.) Such fringes are known as diffraction halo fringes and a fringe pattern obtained in this way is shown in Figure 3.2(*b*).

It is important to note that if Q_2 is less than half the minimum grating spacing, an equation of the form of (3.1) cannot be satisfied, i.e. component gratings in the displaced speckle pattern cannot be in anti-phase with equivalent gratings in the original pattern, so that halo fringes are not observed. This imposes a lower limit to the magnitude of the displacement which can be measured with a given optical configuration. In practice one requires two minima of intensity to be present in the diffraction halo if an accurate measurement of the displacement is to be made. For this to occur Q must be greater than or equal to minimum grating spacing. The latter will correspond to the Airy disc diameter in the speckle pattern recording plane and has a value of $\sim 1.22f\lambda/a$ where f is the focal length of the viewing lens and a its aperture (Section 1.6.2).

3.2.2 *Point-by-point surface displacement measurement using speckle pattern photography*

It is assumed in arriving at equations (3.1) and (3.7) that the displacement $Q_2 = md_2$ is constant over the area of the speckle photograph which is illuminated. If Q_2 varies across the illuminated area, the cancellation of the two component gratings will be incomplete and it is seen that when the variation is such that the angular spacing of the halo fringes produced by one part of the plate is twice that of the halo fringes produced by another part, the fringes will tend to cancel one another out.

Let the displacement at the edge of the illumination area be $Q_2 + \Delta Q_2$; then the halo fringe spacing in the light diffracted from this area is given from equation (3.11) by

$$(\alpha + \Delta\alpha) = \frac{\lambda}{Q_2 + \Delta Q_2} \tag{3.12}$$

To avoid cancellation of the halo fringes, we require that

$$\Delta\alpha \ll \alpha$$

which means that

$$\Delta Q_2 \ll Q_2 \tag{3.13}$$

Thus, when Q_2, the speckle displacement in the x_2-direction, varies across the speckle photograph, the latter must be illuminated by a beam whose width is such that equation (3.13) is satisfied; otherwise halo fringes will not be observed. An unexpanded laser beam is often used for this purpose and the diffracted light viewed at a large distance from the plate – see Figure 3.3. A given area of the photograph is illuminated, and the angular spacing of the halo fringes is found. It follows from equation (3.11) that

$$Q_2 = \lambda/\alpha$$

Q_2 can therefore be determined, and since

$$Q_2 = md_2$$

the displacement of the equivalent point in the object may be measured.

When the displacement is in the d_2 direction only, the fringes will lie parallel to the η_3-axis in the Fourier plane as shown for example, in

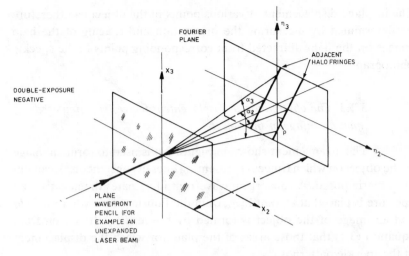

Fig. 3.3 The geometry of diffraction-halo fringes formed as the result of an object displacement d_2, d_3. The speckle displacement over the region of the negative is md_2, md_3, $\tan \rho = d_2/d_3$, $\alpha_2 = \lambda/md_2$ and $\alpha_3 = \lambda/md_3$ (m is the magnification of the object viewing lens)

Figure 3.2(b). Conversely a d_3 displacement ($d_2 = 0$) will result in the fringes lying parallel to the η_2-axes. For the case where the displacement has components

$$Q_2 = md_2$$

and

$$Q_3 = md_3$$

the in-plane displacement vector lies at an angle ρ to the x_3-axis where

$$\tan \rho = \frac{Q_2}{Q_3} = \frac{d_2}{d_3} \tag{3.14}$$

Thus the resultant halo fringe will lie at the same angle ρ to the η_3-axes as indicated in Figure 3.3. Furthermore, the fringe separations in the halo measured parallel to the η_2- and η_3-axes respectively will have angular components

$$\alpha_2 = \lambda/Q_2 \tag{3.15a}$$
$$\alpha_3 = \lambda/Q_3 \tag{3.15b}$$

The in-plane displacement at various points in the object can therefore be determined by measuring the orientation and spacing of the halo fringes on the light diffracted from corresponding points in the speckle photograph.

3.2.3 *The observation of displacement fringes superimposed upon an image of the object*

The arrangement shown in Figure 3.4 is used to form an image of the object on which fringes representing lines of constant displacement are superimposed. A convergent wavefront illuminates the negative; an aperture is placed at a point $P(\eta_2, 0)$ in the Fourier plane where $\eta_2 = l\alpha$ and an image of the object is formed by the lens L_2. It is seen from equation (3.1) that those areas of the plate for which the displacement of the speckle pattern is

$$Q_2 = (n + \tfrac{1}{2})\lambda/\alpha = (n + \tfrac{1}{2})\lambda l/\eta_2, \qquad n = 0, 1, 2, 3, \ldots \qquad (3.16)$$

form diffraction halo fringes which have minima occurring at $(\eta_2, 0)$ so that the final image contains no light from these regions. Thus, the image has fringes which define constant speckle pattern displacement in the x_2-direction at intervals of

$$\Delta Q_2 = \frac{\lambda l}{2\eta_2} \qquad (3.17)$$

When the only object displacement present is parallel to the x_2-axis, the speckle pattern is displaced solely in the x_2-direction, and the fringes will be parallel to the x_3-axis with a spacing dependent on the rate at which d_2, and hence Q_2, varies with respect to the x_2-coordinate. The

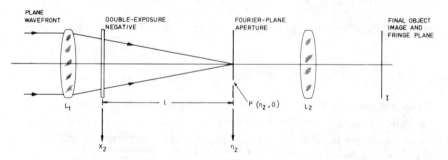

Fig. 3.4 An optical arrangement for the observation of double-exposure SPP displacement fringes superimposed upon an image of the object. The fringe pattern is observed in the image-plane I of the lens L_2

fringe spacing is therefore a function of the displacement gradient of the speckle pattern and hence of the displacement gradient of the object.

If the displacement of the object at a given point has components in the x_2- and the x_3-directions, the displacement of the speckle pattern at a corresponding point in the speckle photograph also has components Q_2, Q_3 in these directions; see equations $(3.15a)$ and $(3.15b)$. However, the positions of these fringes in the Fourier plane along the x_2-axis is determined only by the component of the halo fringe spacing in the x_2-direction, so that fringes are observed in the final image when

$$Q_2 = (n + \tfrac{1}{2}) \frac{\lambda l}{\eta_2}, \qquad n = 0, 1, 2, 3, \dots$$

so that these fringes represent only the components in the x_2-direction of the displacement. Thus the spacings of the fringes in the x_2- and x_3-directions are a function of the gradients of d_2 in the x_2- and x_3-directions respectively. This can be seen by the following analysis, which is essentially the same as the derivation of fringe spacing equations in holographic interferometry (see Section 2.4.2). We define the fringe order number $N(r')$ from equation (3.16) as

$$N(r') = \frac{\eta_2}{l\lambda} Q_2(\mathbf{R}) - \tfrac{1}{2} = \beta Q_2(\mathbf{R}) - \tfrac{1}{2} \qquad (3.18)$$

where $\mathbf{R}$ represents a position in the speckle photograph plane, r' a position in the final image plane, $N(r')$ is not necessarily integral, and $\beta = \eta_2 / l\lambda$.

The fringe order number at a point r' is related to the value at a point r'_0 by the equation

$$N(r') = N(r_0') + \left(\frac{\partial N}{\partial x_i'} \right) (x_i' - x_{0i}') + \cdots \qquad i = 2, 3$$

where x_i' and x_{0i}' are the components of r' and r'_0 respectively. Neglecting higher order terms we have

$$\Delta N = N(r') - N(r_0') = \left(\frac{\partial N}{\partial x_i'} \right) (x_i' - x_{0i}')$$

Now $\Delta N = 1$ when x_i' and x_{0i}' represent equivalent points in adjacent fringes. If $\Delta x_2'$ and $\Delta x_3'$ are the fringe spacings in the x_2- and x_3-directions we can therefore write

$$\Delta x'_{2,3} = \frac{1}{(\partial N / \partial x'_{2,3})} \qquad (3.19)$$

where $(\partial N/\partial x_{2,3})$ is found by differentiating equation (3.18), and equation (3.19) then becomes

$$\Delta x'_{2,3} = \frac{1}{\beta(\partial Q_2/\partial x'_{2,3})} \qquad (3.20)$$

Let the magnification of the final image with respect to the speckle photograph be M, so that

$$x'_{2,3} = Mx_{2,3}$$

This means that when the d_2 displacement gradient of the object in the x_2- and x_3-directions are written in the usual way $d_{2,2}, d_{2,3}$ (Section 2.2.1), equations (3.20) can be expressed in the form

$$\Delta x'_2 = \frac{1}{\beta} \frac{M}{\partial Q_2/\partial x_2} = \frac{l\lambda}{\eta_2} \frac{M}{m} \frac{1}{d_{2,2}} \qquad (3.21a)$$

$$\Delta x'_3 = \frac{1}{\beta} \frac{M}{\partial Q_2/\partial x_3} = \frac{l\lambda}{\eta_2} \frac{M}{m} \frac{1}{d_{2,3}} \qquad (3.21b)$$

since $Q_2 = md_2$. Hence it is seen that measurement of the fringe spacings in the x_2- and x_3-directions in the final image enable the displacement gradients $d_{2,2}, d_{2,3}$ of the object to be found. Furthermore the sensitivity of the measurement can be altered by altering the position of the aperture $(\eta_2, 0)$.

A similar analysis shows that an aperture located at $(0, \eta_3)$ gives fringes which are sensitive to displacements in the x_3-direction only. These correspond to constant speckle displacements Q_3 that result from object displacement d_3 satisfying the equation

$$md_3 = Q_3 = (n + \tfrac{1}{2}) \frac{\lambda l}{\eta_3}, \qquad n = 0, 1, 2, 3, \ldots$$

The fringe spacings in the x_2- and x_3-directions are related to displacement gradients in the x_3-direction by an expression similar to equation (3.20) as

$$\Delta x'_{2,3} = \frac{1}{\beta} \frac{1}{\partial Q_3/\partial x'_{2,3}} \qquad (3.22)$$

This can be written in the form of equation (3.21)

$$\Delta x'_i = \frac{l\lambda}{\eta_3} \frac{M}{m} \frac{1}{d_{3,i}}, \qquad i = 2, 3$$

In the discussion so far we have considered only the case where the speckle pattern has been recorded in a plane for which the object is in focus. In this case, the speckle motion can be considered to be unaffected either by the shape or direction of the illuminating wavefront or by the presence of out-of-plane translation or displacement gradients (tilt) (see however Section 3.5.2). When the speckle pattern is recorded in a plane for which the object is not in focus, as discussed in the following section, the situation is not so straightforward. As will become apparent the spacing of the fringes is dependent upon both the shape of the illumination wavefront and the position of the recording plane. Under specific conditions they can be made sensitive purely to tilt.

3.3 Defocussed plane recording speckle pattern photography

It was seen in the previous section that when the speckle photograph is recorded in the plane in which the object is imaged, fringes can be obtained using either of the viewing techniques discussed in Sections 3.2.2 or 3.2.3 to give fringes which depend upon the absolute in-plane displacement or define contours of constant in-plane displacement i.e. where the fringe spacing is inversely proportional to the in-plane displacement gradient. As will become apparent in this section, the use of any other recording plane makes the motion of the speckle pattern sensitive to out-of-plane as well as in-plane displacements. The speckle motion may however be made sensitive only to out-of-plane displacement gradients by a suitable choice of object illumination geometry and speckle recording plane. It will be shown that the components Q_j, $j = 2, 3$ of the speckle motion are related to the out-of-plane displacement gradient $d_{1,j}$, $j = 2, 3$ by the equations

$$Q_j = Gd_{1,j}, \qquad j = 2, 3 \tag{3.23}$$

where G is a constant which depends on the viewing and illumination geometry.

The point by point measurement technique described in Section 3.2.2 can be used to find Q_2 and Q_3 (see equations 3.15a and 3.15b) and $d_{1,2}$, $d_{1,3}$ can then be evaluated for various points on the object surface.

Alternatively, the technique discussed in Section 3.2.3 can be used to form an image of the speckle photograph giving fringes that define contours of constant Q_2 and Q_3. Equations similar to (3.21) and (3.22) can then be used to evaluate $(\partial Q_j/\partial x_k)$, $j = 2, 3$, $k = 2, 3$. We have

$$\frac{\partial Q_2}{\partial x_2} = G \frac{\partial (d_{1,2})}{\partial x_2} \tag{3.24a}$$

$$\frac{\partial Q_2}{\partial x_3} = G \frac{\partial (d_{1,2})}{\partial x_3} \qquad (3.24b)$$

$$\frac{\partial Q_3}{\partial x_2} = G \frac{\partial (d_{1,3})}{\partial x_2} \qquad (3.24c)$$

$$\frac{\partial Q_3}{\partial x_3} = G \frac{\partial (d_{1,3})}{\partial x_3} \qquad (3.24d)$$

These equations in conjunction with equations (3.20) and (3.22) enable the first derivatives of the displacement gradients to be found from the spacings of the fringes obtained by placing apertures at $(\eta_2, 0)$ and $(0, \eta_3)$ respectively in the Fourier viewing plane.

In Section 3.3.1 it is demonstrated that when the object is illuminated by a plane wavefront and the recording plane is located at the focal plane of the lens, the speckle motion is sensitive only to out-of-plane displacement gradients. The value of G for this arrangement is shown to be $(1 + \cos \theta)$ where θ is the angle of incidence of the illumination wavefront to the surface normal.

In Section 3.3.2, an arrangement is discussed which gives speckle motion in the recording plane sensitive only to out-of-plane tilt when the object is illuminated by a diverging wave. This arrangement is more convenient when large objects are being observed since it is not necessary to employ large collimating lenses.

3.3.1 *Defocussed speckle pattern photography using collimated object-beam illumination and focal plane recording*

The surface D, Figure 3.5, is illuminated by a plane wavefront U_0 which is incident at an angle θ to the surface normal. The photographic plate is located in the focal plane of the lens L which views the surface normally. (This arrangement has been discussed in Section 1.8.3.) The light illuminating the point Q in the recording plane will consist of components of light scattered from the area A centred around P which travel in a direction parallel to PO, where O is the centre of the lens. The area A is the projection of the lens aperture onto the object plane and when the width of the object area illuminated is considerably less than the object to lens distance, the width of A is approximately equal to a, the viewing lens aperture diameter. The phases of the components contributing to the light at Q will vary as the surface height varies, and the resultant amplitude and hence, intensity, of the light will vary randomly across the recording plane – i.e. a speckle pattern is obtained.

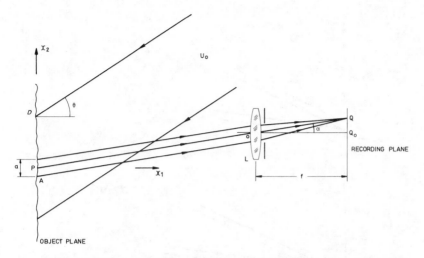

Fig. 3.5 Defocussed speckle pattern recording. In this arrangement the speckle pattern is recorded in the focal plane (or Fourier plane) of the lens L, focal length f

When the object is rotated, the phase of the component of light scattered from a given point on the object to a given point in the image plane alters. Consider the element A shown in Figure 3.6 which is rotated by an angle $\gamma(\equiv d_{1,2})$. It is seen that the light scattered from a point R_1 at the top of the unrotated element travels a shorter distance than the light scattered from the equivalent point R_1' in the rotated element. The light scattered from the point R_2 at the bottom of the unrotated element travels a longer distance than the light scattered from the equivalent point R_2' in the rotated element. Thus, the relative phases of the components of light arriving at Q in the recording plane alter when the object is rotated.

It will now be shown, however, that the relative phase of the light scattered from R_1' and R_2' to a particular point Q' which is adjacent to Q is the same as the relative phase of the light scattered from R_1 and R_2 to Q so that the intensity of the light scattered to Q' from the rotated object is the same as that scattered to Q from the unrotated object. The speckle pattern is therefore displaced by a distance $QQ'(\equiv Q_2)$ in the recording plane.

Consider the undisplaced element A. The path difference between the rays arriving at R_1 and R_2 is given by

$$\Delta l_1 = a \sin \theta \qquad\qquad (3.25a)$$

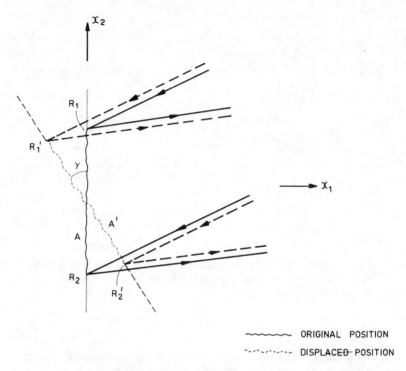

ORIGINAL POSITION

DISPLACED POSITION

Fig. 3.6 Ray paths for the light scattered from the undisplaced element A and displaced element A' viewed and illuminated as shown in Figure 3.5

The path difference between the rays scattered from R_1 and R_2 to Q is given by

$$\Delta l_2 = a \sin \alpha \qquad (3.25b)$$

where

$$f\alpha = QQ_0 \text{ for } \alpha \text{ small, Figure 3.5}$$

Consider the rotated element A'. The path difference between the rays arriving at R_1' and R_2' is now given by

$$\Delta l_1' = a \sin (\theta + \gamma) \qquad (3.26a)$$

The path difference between the rays scattered from R_1' and R_2' to a point Q' which is adjacent to Q is

$$\Delta l_2' = a \sin (\alpha' + \gamma) \qquad (3.26b)$$

where, similarly,

$$f\alpha' = Q'Q_0$$

The relative phase difference from R_1 and R_2 to Q is the same as that from R'_1 and R'_2 to Q' if

$$\Delta l_1 + \Delta l_2 = \Delta l'_1 + \Delta l'_2 \tag{3.27}$$

i.e.

$$\sin \theta + \sin \alpha = \sin (\theta + \gamma) + \sin (\alpha' + \gamma) \tag{3.28}$$

If α, α' and γ are small, we obtain

$$(\alpha - \alpha') = \gamma(1 + \cos \theta)$$

Thus the displacement QQ' is given by

$$QQ' = (\alpha - \alpha')f = f(1 + \cos \theta)\gamma \tag{3.29}$$

Thus, a tilt of the object $d_{1,2}$ will give a shift of the speckle pattern in the x_2-direction of

$$Q_2 = f(1 + \cos \theta)d_{1,2} \tag{3.30a}$$

and a tilt $d_{1,3}$ gives a shift in the x_3-direction of

$$Q_3 = f(1 + \cos \theta)d_{1,3} \tag{3.30b}$$

A tilt having components in both the x_2- and x_3-directions results in fringes that have angular spacing components (Figure 3.3) of

$$\alpha_2 = \frac{\lambda}{f(1 + \cos \theta)d_{1,2}} \tag{3.31a}$$

$$\alpha_3 = \frac{\lambda}{f(1 + \cos \theta)d_{1,3}} \tag{3.31b}$$

These are oriented at an angle ρ to the α_3 Fourier-plane axes where

$$\tan \rho = \frac{d_{1,3}}{d_{1,2}} \tag{3.32}$$

The doubly exposed speckle photograph can thus be used to find the out-of-plane displacement gradients or their first derivatives using the viewing methods discussed in Sections 3.2.2 and 3.2.3 respectively.

In arriving at equation (3.29) it was assumed that the area A' in the rotated object which illuminates the point Q' corresponds exactly to the area A in the unrotated object which illuminates Q. This condition can

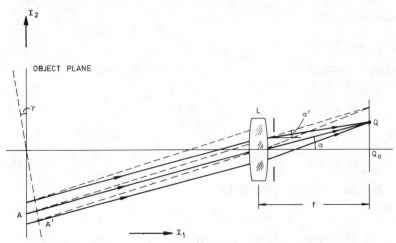

Fig. 3.7 The aperturing effect in defocussed speckle pattern
photography that results from a change in the direction of scattering

never be exactly satisfied. Consider for example the object D which is
rotated about its centre (see Figure 3.7). The area A illuminates the
point Q in the image plane. The equivalent element A' in the rotated
object surface must satisfy equation (3.27) to give a displaced speckle
pattern. However the light scattered from A' in the direction α' is
partially obscured by the lens aperture, and light from adjacent areas
travelling in the same direction is transmitted to P'. Thus the speckle
patterns are not identical and the fringe visibility is reduced. Provided
that the fraction of the light from A' which is obscured is considerably
less than the fraction transmitted, the effect will be small.

When the recording plane is in the focal plane of the lens, rigid body
in-plane displacements, displacement gradients and out-of-plane rigid
body translations decorrelate the speckle pattern but do not significantly
alter its position. For example, when the element A is displaced in its
plane, the optical path difference between the rays arriving at the
displacement points R_1', R_2' is the same as for the undisplaced element.
Also the path difference from $R_1'R_2'$ is the same as from R_1R_2 to Q.
Because the object is shifted, some of the light from A' is obscured by
the lens aperture, and the speckle pattern becomes partially decorrelated.

3.3.2 *Tilt sensitive defocussed speckle pattern photography using diverging object-beam illumination*

It can be shown that if the object is illuminated by a diverging
wave, the speckle pattern recorded in a general plane is displaced when

the object undergoes an in-plane displacement, and also when it under-goes an out-of-plane tilt. The fringe pattern obtained from such a speckle photograph is not then uniquely related to a particular form of object displacement and cannot be used to find the displacement.

Gregory (9) has demonstrated that when the object is illuminated by a divergent beam whose point of divergence is located at a distance r from the object surface, and the speckle photograph is recorded in a plane for which the lens is focussed on a plane at a distance r behind the object surface (see Figure 3.8), then the motion of the speckle pattern is a function only of surface tilts. The derivation of this result, though not intrinsically difficult, is rather complicated and for that reason it is given in Appendix D.

It is found that the angle by which the viewing direction must be altered in order to keep the speckle pattern intensity unchanged is

$$(\alpha - \alpha') = \frac{(1 + \cos \theta)\gamma}{(r + l_1)} \tag{3.33}$$

where l_1 is the object to lens distance, θ is the angle of illumination and γ is the angle of surface tilt. The displacement Q of the speckle pattern is then given by

$$Q = \frac{r(1 + \cos \theta)f\gamma}{(l_1 + r - f)} \tag{3.34}$$

The magnitude and direction of Q can be measured using one of the viewing techniques discussed in 3.2.2 and 3.2.3. Values of $d_{1,2}$ and $d_{1,3}$ may then be determined; the geometric factor G is given by

$$G = \frac{r(1 + \cos \theta)f}{(l_1 + r - f)} \tag{3.35}$$

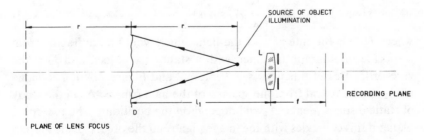

Fig. 3.8 The defocussed-plane recording geometry for the measurement of surface tilt when diverging object-beam illumination is used (after Gregory (9))

3.4 The measurement of dynamic displacements using speckle pattern photography

In Section 2.6 three basic techniques were discussed which enabled dynamic displacements to be measured using holographic interferometry. Similar techniques can be used in speckle pattern photography and are as follows:

(i) time-averaged speckle pattern photography (Section 3.4.1);
(ii) stroboscopic speckle pattern photography (Section 3.4.2);
(iii) dual pulsed speckle pattern photography (Section 3.4.3).

The first two techniques are usually employed to measure periodic displacements. The last can be used to measure displacements occurring in a very short time, for example transients and sections of large-amplitude periodic displacements.

3.4.1 *Time-averaged speckle pattern photography*

Consider first the speckle photograph recorded from an object which is vibrating periodically. It was shown in Section 3.2.1 that the intensity in the recording plane could be described by the convolution of a two-point Delta function with the original speckle pattern intensity distribution. When the photographic recording time is greater than the period of the object vibration, the intensity at a point in the image plane can analogously be described by the convolution of the probability distribution of the position of a point in the object with the stationary object intensity distribution.

It may be shown (13) that when the speckle displacement is described by simple harmonic oscillation of amplitude A_2 in the x_2-direction, the distribution of light in the Fourier plane is given by $I_{TA}(\alpha)$

$$I_{TA}(\alpha) = I(\alpha)J_0{}^2\left(\frac{2\pi}{\lambda}A_2\sin\alpha\right) \tag{3.36}$$

where $I(\alpha)$ is the intensity of the light which would be diffracted from a speckle photograph obtained with a stationary object, and J_0 is the zero order Bessel function (see Table 2.2 and Figure 2.26). The value of A_2 can be found from the spacing of the halo fringes. As in the study of static displacements, A_2 will depend on the position of the recording plane; if this coincides with the image plane and the object-plane motion has a component of in-plane harmonic oscillation of amplitude $(a_0)_2$ where

$$d_2(t) = (a_0)_2\sin\omega t,$$

then

$$A_2 = M(a_0)_2 \tag{3.37}$$

If a defocussed recording plane is used and an out-of-plane harmonic oscillation of amplitude $(a_0)_1$ exists such that

$$d_1(t) = (a_0)_1 \sin \omega t$$

then

$$\frac{\partial d_1(t)}{\partial x_2} = \frac{\partial (a_0)_1}{\partial x_2} \sin \omega t$$

and

$$A_2 = G \frac{\partial (a_0)_1}{\partial x_2} \tag{3.38}$$

where G is the geometrical function given by equations (3.30) and (3.35).

Thus, time-averaged speckle photography may be used to determine independently the amplitude of either in-plane or out-of-plane harmonic oscillations.

When fringes superimposed upon the object image are obtained (Section 3.2.3), these will represent lines of constant A_2 with an intensity distribution corresponding to that of equation (3.36). Note that the J_0^2 distribution will have maximum intensity when

$$\frac{2\pi}{\lambda} A_2 \eta_2 = n; \qquad n = 0, 3.8, 7.0, \ldots$$

Therefore, when the Fourier plane aperture is placed at the coordinate $(\eta_2, 0)$, the bright fringes in the final fringe pattern will correspond to contours of constant vibrational amplitude for which

$$A_2 = n \frac{l\lambda}{\eta_2} \tag{3.39}$$

In the above equation A_2 depends, as usual, upon the initial plane in which the speckle pattern was recorded.

It should be noted that since the distribution of the intensity in the diffraction halo is the Fourier transform of the probability distribution of the speckle position, the form of the object motion can be also found from this distribution (13). One can therefore distinguish between harmonic oscillations, linear motions etc.

3.4.2 *Stroboscopic speckle pattern photography*

In stroboscopic speckle photography, a continuous output laser is modulated so that two pulses are produced during each cycle of the object vibration. The pulses are made to occur at specific points in the vibration cycle as shown, for example, in Figure 4.11; the photograph will then record the positions of the speckle pattern at these two points in the vibration cycle. Consequently the diffraction halo will contain $\cos^2$ fringes whose spacing enables the speckle displacement to be measured. The object in-plane displacement or tilt can then be calculated using the same equations derived for static displacements (Sections 3.2 and 3.3).

When the position and separation of the pulse is varied, the variation of the object displacement throughout the vibration cycle can be determined, though such a measurement process may be extremely tedious.

3.4.3 *Dual pulsed speckle pattern photography*

A dual pulsed laser of the type used in dual pulsed holographic interferometry (Section 2.6.3) can be used to record two speckle pattern photographs of a moving object in rapid succession, i.e. separated by typically 10 to 10^2 µs. The resultant halo fringes (or fringes superimposed upon the object image) can be used to measure the displacement between exposures in the usual way (Sections 3.2 and 3.3).

3.5 Limiting factors in speckle pattern photography

The factors which limit the measurement of displacements by speckle pattern photography have been considered in references (7), (14) and (15) and are briefly discussed here.

3.5.1 *Image formation considerations*

The relationship between the motion of the speckle pattern in the recording plane and the object motion is determined by the location of the recording plane; hence, errors in the interpretation of the fringes will arise when the photographic plate is incorrectly located and when the focus varies across the field of view. The latter will occur when, for instance, a flat object is imaged by a large aperture lens since the image will be curved. Consequently, parts of the speckle pattern in the photographic plate will be slightly defocussed and will be sensitive to tilt as well as to in-plane displacement.

These effects are minimized by using a long focal length lens, so that only a narrow angular field of view is employed (14).

3.5.2 Sensitivity

To observe fringes in a speckle photographic system, the displacement of the object must be such that the displacement of the speckle pattern in the recording plane is greater than the speckle size in that plane (Section 3.2.2); thus the minimum in-plane displacement or out-of-plane tilt which can be detected is determined by the speckle size.

For in-plane measurement, the minimum displacement which will give rise to fringes is given by

$$d_{\min} > q_s/m \qquad (3.40a)$$

where q_s is the speckle size and m is the magnification of the viewing system. In order to minimize lens aberrations and hence, error fringes, the magnification should be unity. In this case, $d_{\min}$ is given by

$$d_{\min} > \tfrac{1}{2}\lambda(f/a) \qquad (3.40b)$$

For a lens having a numerical aperture of 2, $d_{\min} = 0.5$ μm. When the object is viewed with fringes superimposed (Section 3.2.3), the fringe position can be estimated to an accuracy of about one-fifth of a fringe, so that it should be possible to measure displacements to an accuracy of 0.1 μm. It has been shown (14), however, that when the resolution-element diameter in the object is this small, object tilt and out-of-plane displacement give rise to apparent in-plane displacements so that errors occur except when the motion is purely in-plane.

Ennos shows (15) that a sensitivity to tilt as high as that of holographic or speckle correlation interferometry can be obtained, but errors are likely to be introduced as a result of lens aberration and incorrect location of the recording plane. The latter are minimized if minimum image demagnifications are used.

The total number of fringes which can be observed is limited by the speckle size (see Appendix F). It has been shown in the latter that the ratio of fringe spacing to speckle size must be greater than 5 for fringes to be observed; the maximum displacement or tilt which can be observed is thus limited by this factor.

In principle the sensitivity can be reduced to any required value by decreasing the magnification of the viewing system; with a given viewing lens this requires that the object to lens distance is increased. However, as the magnification is decreased, the image plane moves nearer to the focal plane, and errors due to mislocation of the recording plane are more likely to occur.

When the recording plane is located to give sensitivity to one form of motion, other motions will in general tend to decorrelate the speckle pattern (see Appendix F) reducing the visibility of the fringes.

3.5.3 *Object size*

The maximum area which can be inspected in one view is limited only by the laser power available; however, when a large object is imaged onto a small area, errors due to lens aberration may again arise. Gregory (16) (183–223) has used the technique extensively in testing structures such as pressure vessels, wing tips and antenna dishes inspecting areas of up to 1 m^2 in one view.

Very small areas can also be observed. Luxmoore (17) has inspected crack tips on areas of 1 mm^2. Such measurements are however quite difficult; Luxmoore used an x–y vernier traverse table to compensate for rigid body movements, and the measurements entailed a lengthy calibration procedure for each specimen.

3.6 Speckle pattern correlation interferometry

In this section the basic principle of correlation fringe formation is explained in terms of the fringes obtained using an out-of-plane displacement sensitive interferometer. Two methods of observing the fringes are discussed. In Section 3.6.1, the interferometer is described and the relationship between the correlation of the two speckle patterns and the phase change produced by the displacement is derived. In 3.6.2 a 'live' technique of displaying the correlation in the form of a fringe pattern is discussed, and in 3.6.3, a double exposure method requiring the use of a Fourier viewing technique is given. (It should be noted that Electronic Speckle Pattern Interferometry, which is discussed in Chapter 4, provides another way of observing correlation fringes.)

3.6.1 *The principle of speckle pattern correlation fringe formation*

Consider the interferometer shown in Figure 3.9. A plane wavefront U_0 is split into two components of equal intensity by the beamsplitter B. These illuminate the optically rough surfaces D$_1$ and D$_2$. (This is a similar arrangement to that of the Michelson interferometer shown in

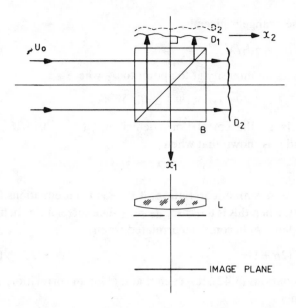

Fig. 3.9 The Michelson arrangement of out-of-plane displacement
sensitive speckle pattern correlation interferometer

Figure 1.15 but the mirrors have been replaced by non-specular surfaces.) These wavefronts scattered from D_1 and D_2 interfere on recom-
bination at B and are recorded in the image plane of the lens–aperture
combination L. The intensity distribution in that plane will consist of
the interference pattern formed between the image-plane speckle patterns of D_1 and D_2 as 'seen' in the dashed position. Let $U_1 = u_1 \exp i\psi_1$
and $U_2 = u_2 \exp i\psi_2$ be the complex amplitudes of these wavefronts
where u_1, u_2 and ψ_1, ψ_2 correspond respectively to the randomly varying
amplitude and phase of the individual image plane speckles. The intensity
of a given point in the image plane will be $\mathscr{I}_1$ where

$$\mathscr{I}_1 = I_1 + I_2 + 2\sqrt{I_1 I_2} \cos \Psi \tag{3.41}$$

and

$$I_1 = U_1 U_1^*$$

$$I_2 = U_2 U_2^*$$

$$\Psi = \psi_1 - \psi_2$$

When D_1 is displaced a distance d_1 parallel to the surface-normal the resultant phase change is given by

$$\Delta\phi(d_1) = 4\pi d_1/\lambda \tag{3.42}$$

This will change the intensity at the point to $\mathscr{I}_2$ where

$$\mathscr{I}_2 = I_1 + I_2 + 2\sqrt{I_1 I_2} \cos(\Psi + \Delta\phi(d_1)) \tag{3.43}$$

In Appendix E, the correlation coefficient, $\rho(\mathscr{I}_1, \mathscr{I}_2)$ of $\mathscr{I}_1, \mathscr{I}_2$ is calculated, and it is shown that when

$$\Delta\phi = 2n\pi \tag{3.44}$$

$\mathscr{I}_1$ and $\mathscr{I}_2$ have maximum correlation; it is seen from equations (3.41) and (3.42) that, when this is the case, $\mathscr{I}_1 = \mathscr{I}_2$. The correlation coefficient is zero, i.e. $\mathscr{I}_1$ and $\mathscr{I}_2$ become uncorrelated when

$$\Delta\phi = (2n+1)\pi \tag{3.45}$$

Thus, using equations (3.42), it is seen that maximum correlation occurs along lines where

$$d_1 = \tfrac{1}{2}n\lambda \tag{3.46a}$$

and minimum correlation exists where

$$d_1 = \tfrac{1}{2}(n + \tfrac{1}{2})\lambda \tag{3.46b}$$

so that the variation in correlation represents the variation in d_1, the normal displacement of the object surface. Variations in speckle pattern correlation of this form can be made to appear as a fringe pattern when the techniques discussed in the following two sections (and Chapter 4) are used.

3.6.2 *Live correlation fringe observation*

A photographic recording of $\mathscr{I}_1$ is made, and the developed negative is accurately relocated in its original position (see Section 6.4.2). The object is now displaced so that the light incident on the plate is given by $\mathscr{I}_2$. Because the photograph is a negative, dark areas will correspond to bright speckles in $\mathscr{I}_1$ and light areas to dark speckles. Thus, very little light will be transmitted in those areas were $\mathscr{I}_1 = \mathscr{I}_2$ (maximum correlation) but where this is not the case, the transmission will increase. It is found that the maximum transmission occurs when $\Delta\phi = (2n+1)\pi$, i.e. when the correlation is a minimum. Thus the variation in correlation is shown as a variation in the transmission of the light through the negative.

The visibility of the fringes obtained using this technique is, however, rather low. This is seen as follows:

Let T be the transmittance of the original negative (Section 6.3.1) so that

$$T = 1 - \alpha t \mathscr{I}_1 \qquad (3.47)$$

where

t = exposure time

and

α = gradient of the photographic emulsion transmission/ exposure characteristics (for example, Figure 6.1).

The light transmitted by this negative, I_T, when illuminated by $\mathscr{I}_2$ will be $T\mathscr{I}_2$ and it follows from equation (3.47) that

$$I_T = \mathscr{I}_2 - K\mathscr{I}_1\mathscr{I}_2 \qquad (3.48a)$$

where $K = \alpha t$. The absorbed intensity I_A is therefore given by

$$I_A = K\mathscr{I}_1\mathscr{I}_2 \qquad (3.48b)$$

The mean value of I_A, averaged over many speckles along a line of constant $\Delta\phi$ will define the light absorption for this part of the fringe pattern. Furthermore, where

$$\langle I_A \rangle = K \langle \mathscr{I}_1 \mathscr{I}_2 \rangle \qquad (3.49)$$

(where $\langle \rangle$ indicates a mean value, as before), $\langle \mathscr{I}_1 \mathscr{I}_2 \rangle$ will be of the form

$$\langle \mathscr{I}_1 \mathscr{I}_2 \rangle = \int_0^\infty \int_0^\infty \int_0^{2\pi} p(I_1, I_2, \Psi) \mathscr{I}_1 \mathscr{I}_2 \, dI_1 \, dI_2 \, d\Psi \qquad (3.50)$$

and $p(I_1, I_2, \Psi)$ is the probability distribution of I_1, I_2 and Ψ. The probability distributions of I_1, I_2 and Ψ are statistically independent and this makes the evaluation of the above integral relatively straightforward. It may be shown that maximum visibility fringes are obtained when the mean transmittance of the negative is 0.5 (i.e. $K\langle \mathscr{I}_1 \rangle = 0.5$) giving

$$\langle I_A \rangle = 2\langle I_0 \rangle^2 [3 + \cos \Delta\phi(d_1)] \qquad (3.51)$$

where $\langle I_0 \rangle$ is the mean speckle intensity. The visibility of the fringes is

$$V = \tfrac{1}{3} \qquad (3.52)$$

(see Section 1.4.2). The fringes shown in Figure 3.10 are characteristic of the technique.

Fig. 3.10 Live speckle pattern correlation fringes. These fringes were obtained using the in-plane displacement sensitive interferometer (Section 3.7.3) and define contours of constant d_2 displacement around a hole in a tensile test specimen deformed by a uniaxial force acting parallel to the x_2-axis. (Reproduced by kind permission of Dr David Rowley)

3.6.3 *The Fourier viewing of speckle pattern correlation fringes*

The double-exposure technique for SPC interferometry is based on a Fourier viewing method. Recordings of the object in its original and displaced position are made on the same plate, but with the deliberate introduction of a lateral image shift between exposures. (This can be done in the manner described at the end of this section.) The latter is uniform over the image and sufficient to form typically up to two fringes in the diffraction halo. Correlated regions of the plate for which $\mathscr{I}_1 = \mathscr{I}_2$ will behave in exactly the same way as a singly exposed plate and therefore form a diffraction halo fringe pattern (Section 3.2.1); it will be seen as follows that regions of minimum correlation will not form a fringe pattern in the diffraction halo.

In those regions of the plate where $\mathscr{I}_1 = \mathscr{I}_2$, the intensity at a point in the plate can be expressed in a form similar to that of equation (3.7) as

$$I_{\mathrm{DE}}(X_2) = \int_{-\infty}^{+\infty} \mathscr{I}_1(X_2')[\delta(X_2 - X_2') + \delta(X_2 + \Delta X_2 - X_2')]\, \mathrm{d}X_2' \qquad (3.53)$$

where ΔX_2 represents the lateral image shift.

The intensity of the light diffracted from such regions can be expressed in a form similar to that of equation (3.9) as

$$I_D(\sin \alpha) = I(_{SE}(X_2)) \cdot \cos^2 [(2\pi \Delta X_2/\lambda) \sin \alpha] \qquad (3.54)$$

Thus, the intensity of the light diffracted from these regions falls to zero at angles given by

$$\sin \alpha = \frac{N\lambda}{\Delta X_2} \qquad (3.55)$$

These fringes arise because two identical speckle patterns $\mathcal{I}_1(X_2)$ are displaced with respect to one another. In those parts of the plate where $\mathcal{I}_1(X_2)$ and $\mathcal{I}_2(X_2)$ are uncorrelated, the diffraction halo is that of entirely unrelated speckle patterns. Consequently no halo fringes will be obtained in the light diffracted from such regions. In Figure 3.11 the dashed line

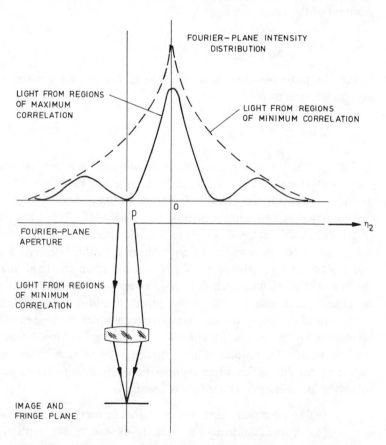

Fig. 3.11 The Fourier filtering arrangement for the observation of double-exposure speckle pattern correlation fringes

therefore represents the distribution of light from uncorrelated regions and the continuous line represents light from regions of maximum correlation. If an aperture is located at a minimum of the halo fringes, *P*, no light is transmitted through the aperture from regions of maximum correlation. The image formed by the lens in Figure 3.11 will have intensity maxima corresponding to lines of minimum correlation, i.e. where

$$\Delta \phi = (2n + 1)\pi \qquad (3.56)$$

since

$$\Delta \phi = 4\pi d_1 / \lambda$$

we see that the fringes represent contours of constant out-of-plane displacement d_1.

An analysis similar to that in Section 3.2.3 shows that the fringe spacing components $\Delta x_{2(3)}$ in the $x_{2(3)}$-directions are related to the gradient of the phase $\Delta \phi$ by

$$\Delta x_i = \frac{2\pi}{\partial(\Delta \phi)/\partial x_i}, \qquad i = 2, 3 \qquad (3.57)$$

For the out-of-plane displacement interferometer discussed in this section we then have

$$\Delta x_i = \frac{\lambda}{2d_{1,i}}, \qquad i = 2, 3 \qquad (3.58)$$

The spatial frequency component of the double exposure negative that results from the image shift causes the negative to act as a diffraction grating. As a result, a diffraction spectrum is obtained when the negative is illuminated by a collimated beam of white light (18). When the viewing of the diffracted light is confined to a single spectral order (i.e. colour) it has the same effect as the Fourier filtering method described above and the correlation fringes are superimposed upon the final image of the object. This viewing technique has two advantages: firstly, it is easier to set up; and, secondly, it does not introduce additional speckle noise.

In both of the above double-exposure techniques the best results will be obtained if the number of fringes in the halo is kept low (ideally two minima should be visible). This requirement can sometimes present experimental difficulties when deformations are being studied and the following experimental procedure is useful:

(*a*) Expose plate with object in undeformed state.
(*b*) Apply deformation and re-expose plate without the introduction of an image shift.

(c) Check if halo fringes can be seen. Sometimes the rigid body motions associated with a deformation are sufficient to form the necessary halo pattern. If this is the case proceed to correlation fringe observation. Alternatively there may be too many fringes or none at all. In such cases it may be possible to estimate the image shift necessary to generate the required halo fringe pattern from the spacing of the fringes observed, if any.

(d) If the halo fringe spacing is not suitable repeat (a) and (b) for the same deformation, introducing the image shift estimated from (c).

(e) Repeat (c).

Stages (d) and (e) may require a few attempts before the correct halo fringe spacing is obtained and it is useful to have a controlled image shift. Suitable lateral translations of the image may be obtained by rotating a glass plate (thickness about 10 mm) in front of the viewing lens. Halo fringe spacings observed in the absence of object deformation can be used to calibrate the plate rotation against the resultant image translation. It will also be appreciated from this discussion that the method may be exploited to observe small surface displacement gradients in the presence of quite large rigid body motions, provided that the deliberate image shift is sufficient to reduce the number of halo fringes to a suitable value.

3.7 Speckle pattern correlation interferometers

In the previous section, an interferometer was described which gives fringes representing lines of constant out-of-plane displacement. The phase difference term $\Delta\phi$ (for example equation 3.42) which gives rise to the variation in correlation of the two speckle patterns may be made sensitive to different components of surface displacement by altering the object and reference beam geometries. Various interferometers are described in this section for which $\Delta\phi$ is a function of out-of-plane displacement, in-plane displacement and out-of-plane displacement gradients. An arrangement is also described which can be used to observe out-of-plane vibration fringes.

3.7.1 *Out-of-plane displacement sensitive interferometers*

The interferometer in Figure 3.9 discussed in Section 3.6.1 gives fringes at intervals of

$$\Delta\phi = 4\pi d_1 / \lambda$$

If the geometry of this interferometer departs from one in which either or both the viewing and illumination directions are no longer parallel to the surface-normal, $\Delta\phi$ has the general form

$$\Delta\phi = \frac{2\pi}{\lambda}(\boldsymbol{n}_{\mathrm{o}} - \boldsymbol{n}_{\mathrm{s}}) \cdot \boldsymbol{d} \tag{3.59}$$

where

$\boldsymbol{n}_{\mathrm{o}}$ = illumination direction

$\boldsymbol{n}_{\mathrm{s}}$ = viewing direction

so that the fringe pattern represents in-plane as well as out-of-plane displacement. (This result follows from the analysis in Section 1.8.3.) Such a situation will arise if divergent illumination is used or the object is non-planar, but it should be noted that as long as the departures from the normal condition are small (typically less than 15°) we can write, to a very good approximation, that

$$\Delta\phi = 2\pi(\cos\theta_1 + \cos\theta_2)d_1/\lambda \tag{3.60}$$

where

θ_1 = angle of object illumination to surface-normal

and

θ_2 = angle of viewing direction to surface-normal

An alternative arrangement which will give the same out-of-plane displacement sensitive fringes is shown in Figure 3.12. Here the object D is illuminated by the object beam U_{o} at an angle θ_1 to the surface normal and an image is formed by the lens L at I. A diverging spherical

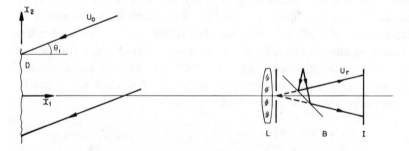

Fig. 3.12 A diagrammatic arrangement of an out-of-plane displacement sensitive speckle pattern correlation interferometer based on a smooth, in-line reference wavefront

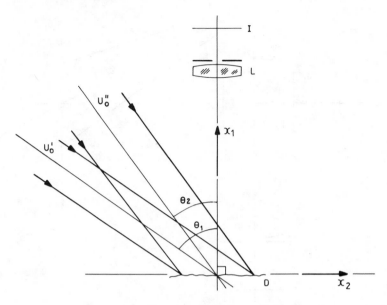

Fig. 3.13 An out-of-plane motion sensitive speckle pattern
correlation interferometer of reduced displacement sensitivity

reference wave U_r is added to the image by the beamsplitter B. When
the object is displaced, the change in the phase of the object beam
relative to the reference beam is given by equation (3.60) with $\theta_2 = 0$,
so that fringes representing out-of-plane displacement are obtained.
(This configuration of interferometer is particularly important in electronic speckle pattern interferometry, Chapter 4.) Interferometers based
on this arrangement may also be used to observe live time-averaged
vibrations. Their principle of operation is described in Section 3.7.4.

Speckle pattern correlation fringes of reduced out-of-plane displacement sensitivity may be obtained using the optical arrangement shown
in Figure 3.13. The correlation phase factor is now given by

$$\Delta\phi(d_1) = \frac{2\pi}{\lambda} d_1(\cos\theta_1 - \cos\theta_2) \qquad (3.61)$$

where θ_1 and θ_2 are the angles of inclination of the two illuminating
wavefronts U_o' and U_o'' to the surface normal. Viewing is in the normal
direction. When θ_1 and θ_2 are approximately equal, $(\cos\theta_1 - \cos\theta_2)$
becomes small and contours of out-of-plane displacement, typically of
the order 10 μm, may be observed. Displacements of this magnitude
rapidly cause speckle pattern decorrelation (Appendix F.2); this limits
the degree of desensitization that can be obtained and limits total

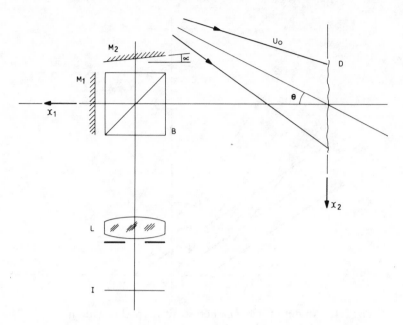

Fig. 3.14 A basic arrangement of speckle pattern shearing
interferometer for the measurement of the first derivative of
displacement gradients

maximum observable displacements that may be observed to typically
10^2 μm.

3.7.2 *Displacement gradient derivative sensitive interferometers*

A form of SPC interferometer (19) which will measure the first
derivative of static displacement gradients is shown in Figure 3.14. The
object D is illuminated by a single wavefront inclined at an angle θ to
the surface-normal and viewed through the Michelson interferometer
arrangement by the lens L. The mirror M_1 lies parallel to the x_2x_3-plane
and M_2 is inclined at a small angle, α, to the x_1-direction. Two images
of D are therefore formed in the plane I as a result of the wavefront
shearing action of this viewing arrangement (see Figure 3.15). (This
arrangement is often referred to as a *speckle pattern shearing interferometer.*) The intensity at a point Q in the image will correspond to
the superposition of the light scattered from two adjacent points in the
original object. Let the separation of these points be S_2 (the shift is
parallel to the x_2-axis) and for simplicity assume an image magnification
of unity – (Figure 3.15).

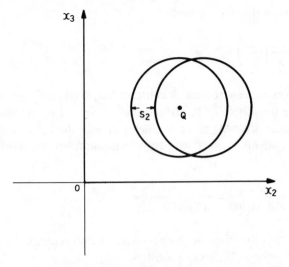

Fig. 3.15 The sheared image geometry obtained using the arrangement shown in Figure 3.14

The object is displaced and d and d' are the respective displacements of the two points illuminating Q. The phase of the light arriving at Q will suffer a phase change of

$$\Delta\phi = \frac{2\pi}{\lambda}(n_\mathrm{o} - n_\mathrm{s}) \cdot (d - d') \tag{3.62}$$

(see equation 3.59), since the two points in the object illuminating the point P are displaced by different amounts. As before (Section 3.7.1), n_o and n_s are the illumination and viewing directions.

When the viewing and illumination directions are normal

$$\Delta\phi = \frac{4\pi}{\lambda}\Delta d_1 \tag{3.63}$$

Now, Δd_1 may be written as

$$d_1(x_2) - d_1(x_{02}) = \frac{\partial d_1}{\partial x_2}\Delta x_2 + \frac{\partial^2 d_1}{\partial x_1^2}\frac{1}{2}(\Delta x_2)^2 \cdots \tag{3.64}$$

where $d_1(x_{02})$ is the displacement at the point x_{02}. The separation S_2 of the two points illuminating a given point in the image plane will generally be sufficiently small that the higher order terms can be neglected in equation (3.64) so that we can write

$$\Delta\phi = \frac{4\pi}{\lambda}\left(\frac{\partial d_1}{\partial x_2}\right)S_2 \tag{3.65}$$

Thus, the fringes represent lines of constant $(\partial d_1/\partial x_2)$.

Clearly, the sensitivity of the fringes can be varied by changing the relative orientations of M_1 and M_2 which will alter the value of S_2.

The fringe spacing in the x_2-direction is found from equation (3.65) and

$$\Delta x_2 = \frac{2\pi}{\partial(\Delta\phi)/\partial x_2} = \frac{\lambda}{2(\partial^2 d_1/\partial x_2^2)S_2} \tag{3.66}$$

Hence, we see that the fringe spacing is constant over regions of constant curvature – i.e. when $\partial^2 d_1/\partial x_2^2$ is constant.

The fringe spacing in the x_3-direction is given by

$$\Delta x_3 = \frac{\lambda}{2(\partial^2 d_1/\partial x_2 \partial x_3)S_2} \tag{3.67}$$

When the image is sheared a distance S_3 in a direction orthogonal to that of S_2 we have

$$\Delta x_2 = \frac{\lambda}{2(\partial^2 d_1/\partial x_2 \partial x_3)S_3} \tag{3.68a}$$

and

$$\Delta x_3 = \frac{\lambda}{2(\partial^2 d_1/\partial x_2{}^2)S_3} \tag{3.68b}$$

This technique is illustrated by the tracings of the interferograms shown in Figures 3.16(a) and (b). The first corresponds to the out-of-plane deflection of a cantilever loaded plate as observed using normal view and illumination holographic interferometry. The fringes define contours of constant d_1 at intervals of $\frac{1}{2}\lambda$. The second interferogram represents the first differential of this pattern, with respect to x_2, obtained in the way described above.

An alternative form of speckle pattern shearing interferometer has been developed by Hung (20). In this arrangement the sheared image is obtained by placing a small-angle wedge over half the viewing lens aperture.

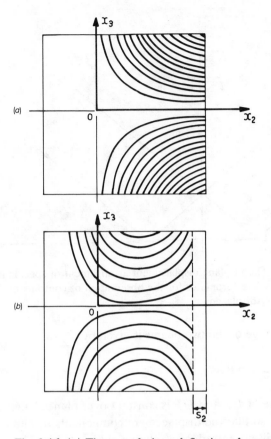

Fig. 3.16 (*a*) The out-of-plane deflection of a cantilever loaded plate. The fringes define contours of constant *d*, at intervals of $\frac{1}{2}\lambda$ and the fringe space $\Delta x_2 \simeq \lambda/2d_{1,2}$. (*b*) The first derivative of the pattern shown in Figure 3.16(*a*). The fringe spacing, Δx_2, equals $\lambda/[2S_2 \partial(d_{1,2})/\partial x_2]$ where S_2 is the magnitude of image shear

3.7.3 *In-plane displacement sensitive interferometers*

The arrangement shown in Figure 3.17 gives fringes which are sensitive to in-plane displacement. Here the object lies in the x_2, x_3 plane and is illuminated by two plane wavefronts, U'_o and U''_o, inclined at equal and opposite angles, θ, to the x_1-axis surface-normal. The positive x_3-axis points into the page and the centre of the viewing lens aperture lies on the x_1-axis. When an element is displaced by a distance $d(d_1, d_2, d_3)$ equations of the form (3.59) can be used to show that the

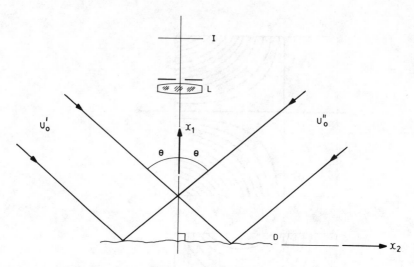

Fig. 3.17 The in-plane displacement arrangement of speckle pattern correlation interferometer for the observation of contours of constant d_2-displacement

relative phase change of the two beams is given by

$$\Delta\phi = \frac{4\pi}{\lambda} d_2 \sin\theta \tag{3.69}$$

The relative phase of U_o' and U_o'' is constant over planes lying parallel to the x_1x_3-plane so that the displacement components d_1 and d_3 lying in those planes will not introduce a relative phase change. This form of interferometer therefore allows in-plane displacement distributions to be observed independently in the presence of out-of-plane displacements (Figure 3.10).

The fringe spacing components $\Delta x_{2(3)}$ in the $x_2(x_3)$-direction are found from equation (3.69) to be:

$$\Delta x_{2(3)} = \frac{2\pi}{\partial(\Delta\phi)/\partial x_{2(3)}} = \frac{\lambda}{2\sin\theta\, d_{2,2}(d_{2,3})} \tag{3.70}$$

An equivalent x_3-axis illumination geometry in which the object is illuminated at equal angles to the x_3-axis will form a fringe pattern where

$$\Delta\phi = \frac{4\pi}{\lambda} d_3 \sin\theta \tag{3.71}$$

and

$$\Delta x_{2(3)} = \frac{\lambda}{2\sin\theta\, d_{3,2}(d_{3,3})} \tag{3.72}$$

It will be seen that fringe spacing measurements must be taken from both illumination geometry fringe patterns in order to determine all the components of the plane strain tensor (Section 2.2.1).

A third in-plane displacement sensitive geometry may be used. In this the object lies in the x_2x_3-plane and is illuminated by plane wave-fronts U'_o, U''_o propagating in the x_2x_1- and x_3x_1-planes at equal angles θ to the surface-normal. Viewing is in the normal x_1-direction. This is referred to as the orthogonal arrangement and it may be shown (21) that

$$\Delta\phi = (d_2 + d_3) \sin \theta \qquad (3.73)$$

and

$$\Delta x_2 = \frac{\lambda}{\sin \theta (d_{2,2} + d_{2,3})} \qquad (3.74a)$$

$$\Delta x_3 = \frac{\lambda}{\sin \theta (d_{3,3} + d_{3,2})} \qquad (3.74b)$$

Under conditions of linear strain i.e. where $d_{2,3} = d_{3,2} = 0$ the ratio of the fringe spacing $\Delta x_3 / \Delta x_2 = d_{3,3}/d_{2,2}$ and hence gives a value for Poisson's ratio.

3.7.4 *The observation of dynamic displacement using speckle pattern correlation interferometers*

Double-exposure SPC techniques based on photographic recording cannot be used for the observation of dynamic displacements in the same way as the SPP method described in Sections 3.4. This is because there is no simple way of introducing the necessary image shift between exposures when time-averaged, stroboscopic or dual pulsed recordings are made. (SPC interferometers could be used for dual pulsed interferometry in the particular case where the intrinsic rigid motions of the object between exposures produced the required halo fringe patterns.) A second factor which limits the use of SPC interferometry is that interferometers in which both the reference and the object beams are in the form of a speckle pattern (for example, Figures 3.13, 3.17) give intrinsically low contrast real-time vibration fringes (Section 4.7.1). For most practical purposes, therefore, SPC interferometers based on the arrangement shown in Figure 3.12, in which the reference beam is smooth, are used for the observation of dynamic displacements. The range of application of this type of interferometer is extended consider-ably when used in electronic speckle pattern interferometry (Chapter

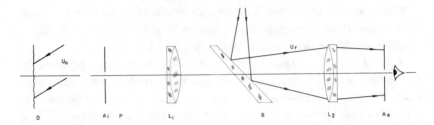

Fig. 3.18 A speckle pattern correlation interferometer for the visual detection of time-averaged vibration modes. (After Archbold *et al* (22) and Stetson (23))

4). In this chapter we will consider only the principle of the technique as used for the visual observation of live time-averaged fringes.

The interferometer in Figure 3.18 was developed by Archbold *et al.* (22) and Stetson (23) and is an extension of the basic arrangement shown in Figure 3.12. It enables the time-averaged fringes to be seen by eye in the image plane of lens L_2. At a given time, t, the intensity in that plane is given by $I(t)$ where

$$I(t) = I_R + I_S + 2\sqrt{I_R I_S} \cos\left[\theta_R - \theta_S + \frac{4\pi}{\lambda} a(t)\right] \quad (3.75)$$

where $a(t)$ represents the position of a given point in the object at time t. The eye or a photographic plate will average the intensity over time τ, to give

$$I_\tau = I_R + I_S + \frac{2}{\tau}\sqrt{I_R I_S} \int_0^\tau \cos\left[\phi_R - \phi_S + \frac{4\pi}{\lambda} a(t)\right] dt$$

When this is evaluated for a sinusoidal vibration,

$$a(t) = a_0 \sin \omega t \qquad \text{(where } 2\pi/\omega \ll \tau\text{)}$$

we find (see Section 2.6.1)

$$I_\tau = \left[I_R + I_S + 2\sqrt{I_R I_S}J_0^2\left(\frac{4\pi}{\lambda} a_0\right) \cos(\phi_R - \phi_S)\right] \quad (3.76)$$

The value of I_τ averaged along many speckles is constant over the whole image, but it can be seen that the contrast of the speckle will vary as the value of the J_0^2 function varies. When J_0^2 has a value of zero, the intensity varies only as I_s varies, whereas when J_0^2 has a maximum value, the intensity varies with variation in I_s and $2\sqrt{I_R I_S} \cos(\phi_R - \phi_S)$.

Thus, the amplitude of the vibration is shown as a variation in speckle contrast, with areas of minimum contrast occurring when

$$a_0 = n\lambda/4, \qquad n = 1, 2, 3, \ldots \qquad (3.77)$$

(see Table 2.2). (Generally, only the zero and first order maxima can be observed. Such a pattern has been photographed by Ek and Molin (24) and shown to be identical in form to that of a time-averaged hologram of the same object.)

3.8 Limiting factors to photographic speckle correlation interferometry

Most of the factors which limit the photographic speckle correlation measurements are the same as those limiting electronic speckle pattern correlation (ESPI) measurements; these are discussed in Section 4.8. In particular the sensitivity for static measurements – Section 4.8.1 – and the decorrelation effects which occur when small areas are viewed are applicable to photographic speckle correlation measurements.

Because photographic SPC is experimentally the most difficult of the techniques discussed here, it is likely to be used only when all the öthers are ruled out. The most likely application is the measurement of in-plane displacements at high sensitivity when an electronic speckle correlation system is not available. (Holographic interferometry enables high sensitivity measurement of out-of-plane motion to be readily made but in-plane motion is detected at slightly lower sensitivity and only then in conjunction with out-of-plane motion – Section 2.5.)

Apart from the observation of time-averaged vibration (Section 3.6.4), dynamic measurements are not possible using photographic SPC interferometry.

The maximum area which can be inspected in one view is mainly limited by the laser power available and the mechanical stability of the system – see Section 6.3.

3.9 Suggestions for further reading

In common with holographic interferometry, speckle pattern interferometry has now been extended and applied to the solution of a wide range of problems. Some of these are outlined below together with references to relevant papers and articles.

The measurement of static and dynamic bending moments using defocussed SPP (American authors in particular often refer to this technique as *speckle pattern shearing photography*): References (25)–(27).

Fluid surface velocity measurement: Reference (28).

The measurement of camera shake: Reference (24).

White light speckle photography: In this technique a special coating that consists of a mixture of small opaque and transparent beads is applied to the object surface. As a result a speckled image of the object is formed when the object is illuminated by white light and viewed with a lens. (The lens has an aperture consisting of equally separated slits whose purpose is to tune the spatial frequency response of the lens to the mean frequency of the surface structure.) Double exposure SPP can then be applied to the study of quite large structures at a reduced displacement sensitivity. The method was developed by Burch and Forno of the National Physical Laboratory and is now used for routine deformation investigations: References (30)–(31).

Stellar speckle pattern interferometry: This is a method by which diffraction limited resolution of stellar objects such as binary stars may be obtained despite the presence of atmospheric turbulence. The processing of the information is very similar to that used in double exposure SPP: Reference (32); 255–80.

Speckle pattern fringe processing and automated analysis: References (33)–(34).

General reading: References (16), (32), (35).

4

Electronic speckle pattern correlation interferometry

4.1 Introduction

In the last chapter it has been shown how photographic processing may be used to form speckle pattern correlation fringes (Section 3.3). The resolution of the recording medium used for this technique need be only relatively low compared with that required for holography since it is only necessary that the speckle pattern be resolved, and not the very fine fringes formed by the interference of object and holographic reference beams. The minimum speckle size is typically in the range 5 to 100 μm (Section 3.1) so that a standard television camera may be used to record the pattern. Thus video processing may be used to generate correlation fringes equivalent to those obtained photographically. This method is known as *Electronic Speckle Pattern Interferometry* or ESPI and was first demonstrated by Butters and Leendertz (1). Similar work has since been described by Biedermann *et al.* (2) and Løkberg *et al.* (3, 4). The major feature of ESPI is that it enables real-time correlation fringes to be displayed directly upon a television monitor without recourse to any form of photographic processing, plate relocation etc. This comparative ease of operation allows the technique of speckle pattern correlation interferometry to be extended to considerably more complex problems of shape measurement (Chapter 5) and deformation analysis (Chapter 7).

Intensity correlation in ESPI is observed by a process of video signal subtraction or addition. In the subtraction process, the television camera video signal corresponding to the interferometer image plane speckle pattern of the undisplaced object is stored electronically. The object is then displaced and the live video signal, as detected by the television camera, is subtracted from the stored waveform. (Note that the camera must remain in the same position throughout.) The output is then high-pass filtered and rectified (this is explained in the next section) and displayed on a television monitor where the correlation fringes may be observed live. For the addition method, the light fields corresponding to the two states are added at the photocathode of the television camera. The television camera detects the added light intensity and the signal is full-wave rectified and high-pass filtered as in the subtraction process. Again the correlation fringes are observed on the television monitor.

The various forms of interferometer discussed in Chapter 3 may be used with a video system. These are:

(i) out-of-plane displacement sensitive interferometer based on the Michelson arrangement (Section 3.7.1);

(ii) desensitized out-of-plane interferometer (Section 3.7.1);

(iii) out-of-plane gradient sensitive interferometer (Section 3.7.2);

(iv) in-plane sensitive interferometer (Section 3.7.3);

(v) out-of-plane sensitive interferometer using a smooth reference beam (Section 3.7.1).

Of the above the first four rely on beams which are both derived from a rough surface so that they are speckled, whilst the last has a smooth reference beam. It will be seen that the restrictions imposed by the limited spatial resolution and dynamic range of the video system are quite different for the smooth reference beam system (v) than for those interferometers (i–iv) which have two speckled beams.

In practice the in-plane sensitive arrangement (iv) and the out-of-plane sensitive arrangement (v) predominate in ESPI. The out-of-plane sensitive arrangement (v) while requiring some care in setting up, is used because it is more efficient in light usage (see Section 4.4) and also because it can be used to give time averaged vibration fringes. (Recently, however, Slettemoen (10) has devised an interferometer which uses two speckled beams to view vibration fringes.)

4.2 The video system (5)

A video system converts an image which is formed on the face plate of a television camera into an equivalent image on a television monitor screen. This face plate has a photosensitive layer on which an electric charge is produced which is proportional to the image intensity. The face plate is scanned by an electron beam that gives rise to a camera output voltage. (A standard television system is scanned in two interlaced sets of lines, generally 625 in total, and a complete scan is performed at a rate of 25 frames per second.) For standard 2.5 mm diameter tubes the active area of the camera is 12×10 mm, so that the image must be reduced to this size. After amplification and the addition of timing pulses the camera signal is used to modulate an electron beam which scans the screen of a television monitor so that the brightness of the screen is made to vary in the same way as the intensity of the original image varies. Ideally, the brightness of the monitor should vary linearly with the intensity of the original image. The exact relationship between the monitor and original image intensities is a complicated function of the electronic processing as well as the brightness and contrast controls of the television monitor. In the analysis of the video display of speckle correlation fringes, it will be assumed that (*a*) the camera output voltage is linearly proportional to the image intensity and (*b*) that the monitor brightness is proportional to the camera output voltage.

This model gives a good qualitative description of the process, but it should be appreciated that a quantitative description of the visibility of ESPI fringes is not possible with this description. (Given the dependence of the fringe contrast on the setting of the monitor controls such an analysis would not in any case be very useful.)

The spatial resolution and dynamic range of video systems are considerably less than those of photographic and holographic emulsions. These factors affect the design of speckle pattern interferometers which use video systems to display the correlation fringes and are discussed in Sections 4.3 and 4.4.

4.2.1 *Speckle pattern correlation fringe formation by video signal subtraction*

The face plate of the television camera is located in the image plane of the speckle interferometer. Under these conditions the output signal from the television camera, as obtained with the object in its initial state, is recorded on a video store which may be a video tape recorder, disc or solid state store (these are discussed in more detail in Section 4.6). The object is then displaced and the live camera signal is subtracted electronically from the stored signal. Those areas of the two images where the speckle pattern remains correlated will give a resultant signal of zero, while uncorrelated areas will give non-zero signals. We can see this by considering the intensities $\mathscr{I}_1$ and $\mathscr{I}_2$ given by equations (3.41) before and (3.43) after displacement:

$$\mathscr{I}_1 = I_1 + I_2 + 2\sqrt{I_1 I_2} \cos \psi \qquad (4.1a)$$

$$\mathscr{I}_2 = I_1 + I_2 + 2\sqrt{I_1 I_2} \cos{(\psi + \Delta\phi)} \qquad (4.1b)$$

If the output camera signals V_1 and V_2 are proportional to the input image intensities, then the subtracted signal is given by

$$V_S = (V_1 - V_2) \propto (\mathscr{I}_1 - \mathscr{I}_2) = 2\sqrt{I_1 I_2}[\cos\psi - \cos{(\psi + \Delta\phi)}]$$
$$= 4\sqrt{I_1 I_2} \sin{(\psi + \tfrac{1}{2}\Delta\phi)} \sin{\tfrac{1}{2}\Delta\phi} \qquad (4.2)$$

This signal has negative and positive values. The television monitor will, however, display negative-going signals as areas of blackness; to avoid this loss of signal, V_S is rectified before being displayed on the monitor. The brightness on the monitor is then proportional to $|V_S|$, so that we have the brightness B at a given point in the monitor image given by

$$B = 4K[I_1 I_2 \sin^2{(\psi + \tfrac{1}{2}\Delta\phi)} \sin^2{\tfrac{1}{2}\Delta\phi}]^{\frac{1}{2}} \qquad (4.3)$$

where K is a constant.

If the brightness B is averaged along a line of constant $\Delta\phi$, we see that it varies between maximum and minimum values $B_{\max}$ and $B_{\min}$ given by

$$B_{\max} = 2K\sqrt{I_1 I_2}, \quad \Delta\phi = (2n+1)\pi, \qquad n = 0, 1, 2 \qquad (4.4a)$$

$$B_{\min} = 0, \qquad\qquad \Delta\phi = 2n\pi, \qquad n = 0, 1, 2 \qquad (4.4b)$$

High-pass filtering of the signals is found to give improved visibility fringes by removing low frequency noise together with variations in mean speckle intensity and is normally used to enhance the fringe clarity.

Comparison of equations $(4.4a)$ and $(4.4b)$ with equations (3.44) and (3.45) shows that the fringes obtained are identical in form to those obtained using the photographic correlation technique. Subtraction correlation fringes photographed from the monitor defining contours of constant out-of-plane displacement are shown in Figure 4.1. These were obtained using an interferometer with a smooth, in-line reference wavefront (Sections 3.7.1 and 4.5.2), the surface having undergone a slight tilt.

4.2.2 *Speckle pattern correlation fringes formation by video signal addition*

In this method, the camera face plate is again placed in the image plane of the speckle interferometer but the two speckle patterns derived from the object in its two states are added together on the camera face plate. The two images do not need to be superimposed

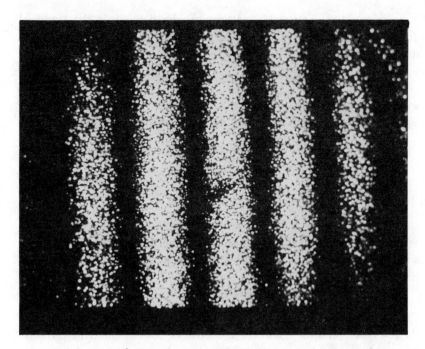

Fig. 4.1 Sin^2 speckle pattern correlation fringes obtained using electronic subtraction. They represent contours of constant out-of-plane displacement and are due to an out-of-plane rotation $d_{1,2}$

simultaneously since a given camera tube has a characteristic persistence time (~0.1 s for a standard tube), so that the camera output voltage will be proportional to the added intensities if the time between the two illuminations is less than the appropriate persistence time. This technique is employed when a dual pulsed laser is used and also for the observation of time-averaged fringes (Section 4.7). Another important application is the observation of two-wavelength shape difference fringes where the object is illuminated simultaneously at two wavelengths (this technique is discussed in Chapter 5).

When the two speckle patterns are added together, areas of maximum correlation have maximum speckle contrast and, as the correlation decreases, the speckle contrast falls. It reduces to a minimum, but non-zero, value where the two patterns are uncorrelated. This is seen as follows:

The voltage V_A is proportional to $\mathscr{I}_1 + \mathscr{I}_2$ and is given by

$$V_A \propto (\mathscr{I}_1 + \mathscr{I}_2) = 2I_1 + 2I_2 + 4\sqrt{I_1 I_2} \cos\left(\psi + \tfrac{1}{2}\Delta\phi\right) \cos \tfrac{1}{2}\Delta\phi$$

$$(4.5)$$

The contrast of the speckle pattern can be defined as the standard deviation of the intensity. For a line of constant $\Delta\phi$, this may be shown to be

$$\sigma_{12} = 2[\sigma_1{}^2 + \sigma_2{}^2 + 8\langle I_1 \rangle \langle I_2 \rangle \cos^2 \tfrac{1}{2}\Delta\phi]^{\frac{1}{2}} \tag{4.6}$$

where σ_1 and σ_2 are the standard deviations of I_1 and I_2. It is seen that σ_{12} varies between maximum and minimum values given by

$$[\sigma_{12}]_{\max} = 2[\sigma_1{}^2 + \sigma_2{}^2 + 2I_1 I_2]^{\frac{1}{2}}, \qquad \Delta\phi = 2n\pi, \qquad n = 0, 1, 2 \tag{4.7a}$$

$$[\sigma_{12}]_{\min} = 2[\sigma_1{}^2 + \sigma_2{}^2]^{\frac{1}{2}}, \qquad \Delta\phi = (2n+1)\pi, \; n = 0, 1, 2 \tag{4.7b}$$

While the contrast of the added intensities varies the mean value along a line of constant $\Delta\phi$ is the same for all $\Delta\phi$, and is given by

$$\langle \mathscr{I}_1 + \mathscr{I}_2 \rangle = 2\langle I_1 \rangle + 2\langle I_2 \rangle \tag{4.8}$$

Thus, when the sum of the two speckle patterns is directly displayed on the television monitor, the average intensity is constant, and the variation in correlation is shown as a variation in the contrast of the speckle pattern but not in its intensity. The DC component of the signal is removed by filtering, and this signal is then rectified. The resulting monitor brightness can then be considered to be proportional to the square root of σ_{12}, so that

$$B = K[\sigma_1{}^2 + \sigma_2{}^2 + 2\langle I_1 \rangle \langle I_2 \rangle \cos^2 \tfrac{1}{2}\Delta\phi]^{\frac{1}{2}} \tag{4.9}$$

Hence the intensity of the monitor image varies between maximum and minimum values given by equations (4.7) and (4.9). Comparison with equations (4.3) and (4.4) shows that the fringe minima obtained with subtraction correspond to fringe maxima obtained using addition. It can also be seen that subtraction fringes have intrinsically better visibility than addition fringes, since the minima have zero intensity, while addition fringes do not. However, when addition is used to observe the fringes, a video store is not required, making the system simpler and less expensive.

If time-averaged vibration fringes are to be observed on the television camera, the period of the vibration must be less than the persistence time τ of the camera. In this case, the intensity is given, for sinusoidal vibration, by equation (3.76) as

$$\mathscr{I} = [I_1 + I_2 + 2\sqrt{I_1 I_2}\, J_0(4\pi a_0/\lambda) \cos \psi] \tag{4.10}$$

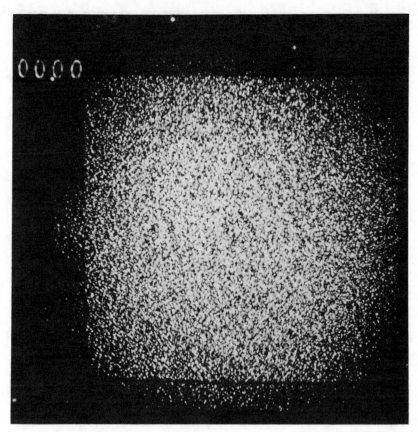

Fig. 4.2 The filtered and rectified image-plane intensity distribution obtained using an interferometer of the type shown in Figure 4.1. The object under investigation was a static flat plate sprayed matt white

When this is displayed directly on the monitor, only a variation in the contrast of the speckle pattern is seen. If, however, it is first high-pass filtered and rectified, the brightness B is given by

$$B = K[\sigma_1^2 + \sigma_2^2 + 2\langle I_1\rangle\langle I_2\rangle J_0^2(4\pi a_0/\lambda)]^{\frac{1}{2}} \qquad (4.11)$$

Correlation fringes are thus observed which map out the variation in a_0, the amplitude of the vibration across the object surface. (The fringe minima correspond to the minima of the Bessel function – see Figure 2.26). A typical result is shown in Figures 4.2 and 4.3. Figure 4.2 shows the filtered and rectified output of the video signal derived from a camera placed in the image-plane of an interferometer of the type shown in

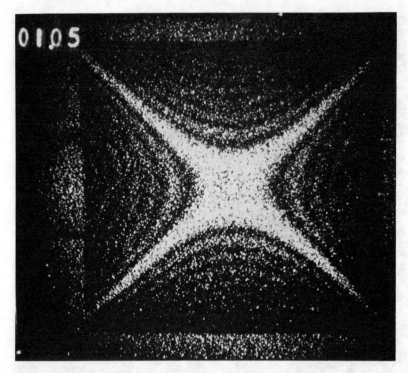

Fig. 4.3 J_0^2 time-averaged fringes observed when the plate in Figure 4.1 was excited at a frequency of 1.05 kHz by a small piezo electric crystal attached to its rear

Figure 4.8. The object under observation is a static flat plate sprayed matt white. Figure 4.3 is the time-averaged correlation fringe pattern that was seen when the same plate was excited at 1.05 kHz. (This corresponds to the fundamental '2D' mode.)

4.3 Spatial resolution of the video system and its effect on speckle pattern correlation measurement

The ability of the video system to resolve fine detail in the image is clearly limited. Thus, if a coarse black and white grid is imaged on the face plate of the television camera, an equivalent grid will be observed on the monitor; as the spacing of the grid is reduced, the contrast of the grid observed on the monitor is reduced, and when the grid spacing is sufficiently small, no grid structure is observed on the monitor.

The limit to the spatial resolution in the vertical direction is governed by the number of scan lines. A standard system having 625 lines and an active face plate area of ~13 × 10 mm should have a minimum vertical resolution of ~20 μm on the face plate. (In practice it is somewhat larger than this.)

The resolution in the horizontal direction is governed by the temporal frequency response of the video system, which is determined by the electronic design of the system (5). Typically, this frequency response is fairly uniform up to 4 MHz and falls to a very low value at 10 MHz. The temporal frequency response is clearly related to the spatial frequency in the image. A sinusoidal intensity distribution in the image plane of spatial frequency $10^6/WNM$ mm^{-1} will give a sinusoidal output voltage of 1 MHz, where

W = width of the active face plate area in mm

N = number of scan lines

M = number of frames/second

For a standard television system, 1 MHz corresponds to a spatial frequency of ~5 mm^{-1} (spacing ~200 μm).

A grid having a 20 μm spacing in the horizontal direction will give an output voltage of frequency 10 MHz. Since the frequency response is very low at 10 MHz, 20 μm can also be taken as the upper limit of the spatial resolution in the horizontal direction.

It was assumed in deriving equations (4.3) and (4.9) which describe the variation in monitor brightness to give subtraction and addition correlation fringes, that the random variations in $\mathscr{I}_1$, $\mathscr{I}_2$ are fully resolved by the television camera, so that the camera signal from a given point in the image is proportional to the intensity at that point – equations (4.2) and (4.5). This is an approximation which is valid when the scale of the fluctuations of $\mathscr{I}_1$, $\mathscr{I}_2$ is greater than the spatial resolution of the camera. When this is not the case, the output voltage is a function of the intensity of the image averaged over the resolution area of the camera, and the effect of this averaging is to reduce the standard deviation or contrast of the output voltage (6). For example, in the case of two speckle patterns whose intensities have the same mean value and standard deviations, one of which is fully resolved by the television camera and the other of which is only partially resolved, the output camera signals will have the same mean values across the image, but the standard deviation (rms value) of the signal will be less for the partially resolved pattern. If the speckles are sufficiently small, the output signal will be uniform. It will be seen below that the effect of this reduction in the standard deviation of the output signal is to reduce the visibility of the correlation fringes.

It is seen from equations (4.3) and (4.4) that the correlation fringes arise from the variation with $\Delta\phi$ of the standard deviation of the subtracted or added signals; if the contrast of the individual signals is reduced, the standard deviation of the subtracted or added signals is also reduced, and the visibility (see equation 1.41) of the correlation fringes is decreased. If we write the output camera signal as

$$V_{T_{1,2}} = V_0 + (V_{corr})_{1,2}$$

where V_0 arises from the $(I_1 + I_2)$ term and $(V_{corr})_{1,2}$ arises from the $2\sqrt{I_1 I_2} \cos \psi$, $2\sqrt{I_1 I_2} \cos (\psi + \Delta\phi)$, terms then it can be seen that, since V_0 remains unchanged when the object is displaced, it is a noise term; consequently the correlation fringe visibility is maximized when the standard deviation of V_{corr} is maximized. We can write the value of V_{corr} at a given point r in the image as

$$(V_{corr})(r)_{1,2} = k\langle\sqrt{I_1 I_2} \cos (\psi + \Delta\phi_{1,2})\rangle \Delta A \qquad (4.12)$$

where ΔA is the resolution area of the camera, k is a constant, $\Delta\phi_1 = 0$, and $\Delta\phi_2 = \Delta\phi$.

If the variation of ψ over the resolution area is very much less than 2π, then the value of V_{corr} for constant $\Delta\phi$ varies between $\pm[k\langle\sqrt{I_1}\sqrt{I_2} \cos \Delta\phi\rangle]_{max}$; the latter quantity is clearly related to the mean values of I_1, I_2, and it will be shown in Section 4.4 that the optimum values for I_1, I_2 are determined by the sort of interferometer used. When the value of ψ varies by 2π or more over the resolution area, then the value of V_{corr} over the whole image is approximately zero, and in this case correlation fringes are not observed.

Thus, the spatial frequency distribution of $\cos \psi$ must be such that the components of that distribution are at least partially resolved by the video system. The restrictions which this condition imposes on speckle correlation interferometers having two speckled beams are discussed in 4.3.1 and on the smooth reference beam interferometer in 4.3.2. Some general conclusions are discussed in 4.3.3.

4.3.1 Spatial resolution: two-speckled-beams interferometers

When the two wavefronts are speckled, the intensities I_1 and I_2 in equations (4.1a, b) are random speckle intensities, and the phase term ψ represents the difference in phase between the two speckle patterns, and may be written as $\psi = \psi_{S1} - \psi_{S2}$. The spatial frequency distribution of ψ will be similar to those of ψ_{S1} and ψ_{S2}, and the maximum spatial frequency f_{max} of the distributions of ψ_{S1} and ψ_{S2} will be determined by the diameter of the viewing lens aperture a, and is given by equation (1.85) as

$$\frac{1}{f_{max}} = \frac{\lambda v}{a}$$

If the object-to-viewing-lens distance is considerably greater than the lens-to-image distance, it can be seen from the lens equation (1.48), that $v \simeq f$. The minimum speckle size g_{min} is then given by

$$g_{min} = \frac{1}{f_{max}} \simeq \lambda \left(\frac{f}{a}\right) = \lambda \, (NA) \tag{4.13}$$

where NA is the numerical aperture of the viewing lens.

If the smallest spot resolved by the video system is 20 μm, corresponding to a spatial frequency of 50 mm^{-1}, then to resolve fully the spatial frequencies arising from ψ_{S1}, ψ_{S2} we must have

$$(NA) < 40 \tag{4.14}$$

Thus, quite a small aperture must be used if the scale of the speckle pattern is to be large enough to be fully resolved by the video.

If the numerical aperture of the lens is reduced below this value, the upper limit of the spatial frequency of the speckle pattern will increase and it will no longer be fully resolved. However, the amount of light transmitted through the lens will increase, so that the value of V_{corr} does not necessarily decrease. An exact analysis of the relationship between V_{corr} and the viewing lens aperture diameter is quite difficult. However, the optimum value can be readily found in practice simply by adjusting the aperture size until maximum speckle contrast is obtained. It is found that a numerical aperture ~ 8 gives good-visibility speckle correlation fringes using the electronic speckle pattern interferometers developed at Loughborough.

4.3.2 Spatial resolution: smooth reference beam system

In this system, I_1 is a random speckle intensity, while I_2 is nominally constant across the image plane. Variations in the value of I_2 will either have very low contrast, or will occur only slowly across the image plane. The phase difference ψ is given by

$$\psi = \psi_S - \psi_T + \frac{2\pi}{\lambda}(l_{OP} - l_{O'P}) \tag{4.15}$$

4 ESPI   176

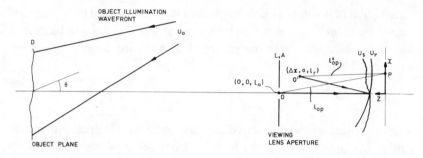

Fig. 4.4 The general geometry of a smooth, on-axis reference beam
interferometer

where O is the position of the centre of the viewing lens, O' is the point
from which the reference beam appears to diverge and P is a point in
the image plane – see Figure 4.4. ψ_T is the phase of the reference beam
at O and is constant across that beam, while ψ_S is the random phase at
O of the light illuminating the point $P(x)$. l_{OP} and $l_{O'P}$ are the distances
from O and O' to P.

It was seen in 4.3 that if ψ varies by 2π or more within the minimum
resolution distance of the camera, the correlation signal V_{corr} will be
very small. The variation in ψ_S is governed by the viewing lens aperture
diameter, as in the case of the two speckled beams system, and a suitable
aperture ensures that it does not vary too rapidly. Clearly, the variation
in the last term $\chi = 2\pi/\lambda\,(l_{OP} - l_{O'P})$ with P must be such that χ varies
by considerably less than 2π over the resolution diameter x_r of the
resolution area.

If O and O' are coincident, χ is zero so that, if possible, this condition
should be satisfied (O and O' are then referred to as being conjugate).

To determine the departure from conjugacy that can be tolerated,
we assume that the coordinates of O and O' are $(0, 0, l_0)$ and $(\Delta x, 0, l_r)$.
When $\Delta x, \Delta l \ll l_o$, where $\Delta l = l_o - l_r$, we have

$$l_{OP} - l_{O'P} \simeq (l_o + x^2/2l_o) - [l_r + (x - \Delta x)^2/2l_r] \qquad (4.16)$$

When second order terms in $\Delta x_r, \Delta l_r$ are neglected, we have

$$l_{OP} - l_{O'P} \simeq \Delta l - \frac{x\,\Delta x}{l_o} - \frac{x^2\Delta l}{l_o^{\,2}}$$

It is seen that a lateral displacement, Δx, of O' with respect to O
produces a linear variation in ψ across the image plane, which varies by

2π across an interval δx given by

$$\delta x = \frac{l_o \lambda}{\Delta x}$$

If this is to be resolved by the video system, δx should be considerably greater than x_r, i.e.

$$\frac{l_o \lambda}{\Delta x} \gg x_r$$

or

$$\Delta x \ll \frac{l_o \lambda}{x_r} \tag{4.17}$$

Thus, the lateral departure from conjugacy must be less than that specified by equation (4.17). For a lens-to-image distance of 100 mm and $x_r = 20\ \mu m$, we have $\Delta x \ll 2.5$ mm.

A longitudinal departure from conjugacy of Δl gives rise to a variation in ψ which varies as x^2. The maximum gradient of ψ occurs at the edge of the image. Here it can be seen that the distance δx in which ψ varies by 2π is given by

$$\delta x = \frac{\lambda l_o^2}{W \Delta l} \tag{4.18}$$

where W is the width of the active area of the camera tube. If this is to be resolved, we must have

$$\frac{\lambda l_o^2}{W \Delta l} \gg x_r$$

or

$$\Delta l \ll \frac{\lambda l_o^2}{W x_r} \tag{4.19}$$

For a lens-to-image distance of 100 mm and $x_r = 20\ \mu m$, we have

$$\Delta l \ll 20\ mm$$

To optimize the visibility of the correlation fringes, Δx_r and Δl should be made as small as possible. It is seen that displacements in the lateral direction give rise to more rapid variation of ψ than equivalent variations in the longitudinal direction so that the lateral shift in particular should be minimized. It is not difficult in practice to do this – see Section 4.5.

In addition, the size of the viewing lens aperture should be adjusted to give optimum speckle contrast.

4.3.3 General conclusions

It is seen from the discussions in the two previous sections that the use of a television camera to observe speckle correlation fringes requires the use of a small aperture in the viewing lens resulting in relatively large speckles. This affects the performance of the interferometer in several ways.

Firstly, the spacing of the correlation fringes must be greater than approximately 1/120th of a screen width. This follows because when the fringe spacing becomes comparable to the minimum speckle size, the fringe visibility decreases and drops to zero when they are equal. This result is derived in Appendix F.1 and is illustrated by the interferogram shown in Figure 4.5. It can be seen that the contrast of these fringes is considerably less than that in Figure 4.1 where the fringe

Fig. 4.5 An interferogram obtained in the same way as that shown in Figure 4.1, but of considerably reduced fringe spacing. This illustrates the reduction in fringe contrast that occurs as the spacing of the fringes approaches the mean speckle size

spacing to speckle size ratio is about 30. A minimum ratio of fringe spacing to speckle size of about five is necessary to give reasonable visibility fringes. Since the minimum speckle size is ~20 μm, this means that the minimum fringe spacing is of the order of 100 μm. This corresponds to a fringe spacing of 1/120th of a face plate of 12 mm width. Hence a maximum of 120 fringes of uniform spacing should, in theory, be observable. With a fringe pattern of variable spacing, the maximum number under optimum viewing conditions is found in practice to be about 50.

Secondly, since the speckles are resolved by the video system, they are seen in the image and give speckled fringes where subjective clarity is less than that of the good quality holographic fringes – compare, for example, Figures 2.8 and 4.1.

Finally, the use of a small aperture means that the system is not very efficient in light usage compared with the equivalent photographic speckle correlation system, expecially in conjunction with the short 'exposure time' (1/25 s) of the video system.

However the speed and convenience of video processing compared with photographic processing generally outweighs these disadvantages.

4.4 The optimization of light intensity levels in Electronic Speckle Pattern Correlation Interferometry

In this section we consider how the maximum image-plane intensity and the relative intensities of reference beam to object beam necessary for optimum interferometer performance are affected by the limited dynamic range of the camera (Section 4.4.1). It will be seen that the type of interferometer, together with optical and electronic noise of the system also have to be taken into account in these calculations (Sections 4.4.2 to 4.4.5). Relatively small departures from optimum operating conditions introduced either by the use of incorrect light levels or by exceeding the resolution limit of the television camera (Section 4.3) can lead to a substantial reduction in fringe contrast in this type of interferometry. These aspects of the system are therefore of particular importance.

4.4.1 *Restrictions imposed by the limited dynamic range of the television camera*

For a given type of camera tube, a certain minimum intensity is required to give rise to a camera voltage which can be detected above the background electronic noise. An increase in the incident intensity gives an increased output camera voltage until the intensity reaches the camera saturation level beyond which the output voltage remains constant for any further increase in incident intensity. The ratio of the minimum to maximum detectable intensities is generally of the order of 10^2 to 10^3.

The intensity of the speckle pattern varies randomly across the image, and to avoid losing information from the speckle pattern, the overall intensity should be below the saturation level of the camera for all, or nearly all, of the picture. If the mean value and the standard deviation of the speckle pattern are given by $\langle I_T \rangle$ and σ_T, then when

$$\langle I_T \rangle + 2\sigma_T < I_{sat} \tag{4.20}$$

the intensity will be less than the saturation level of the camera for 95 % of the image.

If the intensity I_T is given by

$$I_T = I_1 + I_2 + 2\sqrt{I_1 I_2} \cos \psi$$

it can be seen that

$$\langle I_T \rangle = \langle I_1 \rangle + \langle I_2 \rangle$$

and from equation (1.77)

$$\sigma_T{}^2 = \sigma_1{}^2 + \sigma_2{}^2 + 2\langle I_1 \rangle \langle I_2 \rangle$$

Thus if $\langle I_2 \rangle = k\langle I_1 \rangle$, $\gamma_{1,2} = \sigma_{1,2}/\langle I_{1,2} \rangle$ then equation (4.20) can be written as

$$\langle I_1 \rangle [1 + k + 2(\gamma_1{}^2 + k^2 \gamma_2{}^2 + 2k)^{\frac{1}{2}}] < I_{sat} \tag{4.21}$$

It will be seen in the following sections that the value of k which gives optimum fringe contrast is determined by whether the fringes are obtained by addition or subtraction, and also on whether or not a smooth reference beam is used. Furthermore, it is shown that $\langle I_1 \rangle$ has an optimum value which may be determined from equation (4.21) for a given optimum value of k.

4.4.2 *Two-speckled-beams interferometers: subtraction fringes*

When subtraction is used to observe the correlation fringes, the mean intensity of the monitor image along a line of constant $\Delta\phi$ is given from equation (4.3) as

$$B = q|\langle I_1 \rangle \langle I_2 \rangle \sin^2 \Delta\phi|^{\frac{1}{2}} + n_e$$
$$= q\sqrt{k}\langle I_1 \rangle| \sin^2 \Delta\phi|^{\frac{1}{2}} + n_e \tag{4.22}$$

where n_e is the electronic noise of the video system and $q = 4K$. Thus the signal to noise ratio of the fringes may be defined as

$$\text{SNR} = \frac{q\sqrt{k}\langle I_1 \rangle}{n_e}$$

Thus the maximum signal to noise ratio is obtained when $\sqrt{k}\langle I_1 \rangle$ is maximized. The value of this quantity is limited by equation (4.21). If we put

$$\langle I_1 \rangle [1 + k + 2(\gamma_1^2 + k^2\gamma_2^2) + 2k)^{\frac{1}{2}}] = \rho I_{sat}$$

where $\rho < 1$, we than have

$$\text{SNR} = \frac{q\sqrt{k}\rho I_{sat}}{[1 + k + 2(\gamma_1^2 + k^2\gamma_2^2 + 2k)^{\frac{1}{2}}]} \tag{4.23}$$

When two speckled beams are used, we have

$$\gamma_1 = \gamma_2 = 1$$

since $\sigma_{1,2} = \langle I_{1,2} \rangle$. Hence

$$\text{SNR} = \frac{q\rho\sqrt{k}I_{sat}}{3(1 + k)} \tag{4.24}$$

It can be seen that SNR is maximized when $k = 1$, $\rho = 1$.

Thus, when two speckled beams are used in a subtraction correlation system, the mean intensities of the two beams at the image plane should be equal, and the combined peak intensities should be approximately equal to the camera saturation intensity.

4.4.3 *Smooth reference beam interferometers: subtraction fringes*

If I_1 is the speckled beam intensity, and I_2 the smooth reference beam intensity, $\gamma_1 = 1$, but γ_2 is typically $\simeq 1/50$. Evaluation of equation (4.23) shows that in this case SNR has a maximum value when $\rho = 1$ and $k \simeq 2$, but that the variation is relatively insensitive to variations in k and γ_2 when the latter is small. For example

$$\text{SNR}(k = 20) = 0.7 \, \text{SNR}(k = 2)$$

The amount of laser power required to produce a given level of intensity at the image plane from the object is much greater than that required to give the same intensity for a smooth reference beam. If the laser power available is insufficient to set $k = 2$ and $\rho = 1$, a higher value of k should be chosen which allows the camera to operate near saturation level – i.e. $\rho = 1$, $k > 2$.

4.4.4 Two-speckled-beams interferometers: addition fringes

When the fringes are obtained by addition, the mean monitor brightness along a line of constant $\Delta\phi$ is given by

$$B = q\left|\sigma_1{}^2 + \sigma_2{}^2 + 8\langle I_1\rangle\langle I_2\rangle \cos^2 \tfrac{1}{2}\Delta\phi\right|^{\frac{1}{2}} + n_\mathrm{e}$$

which may be written as

$$B = q\langle I_1\rangle\left|1 + k^2 + 8k \cos^2 \tfrac{1}{2}\Delta\phi\right|^{\frac{1}{2}} + n_\mathrm{e} \tag{4.25}$$

The maximum and minimum values are therefore

$$B_\mathrm{max} = q\langle I_1\rangle\left|1 + k^2 + 8k\right|^{\frac{1}{2}} + n_\mathrm{e} \tag{4.26a}$$

$$B_\mathrm{min} = q\langle I_1\rangle\left|1 + k^2\right|^{\frac{1}{2}} + n_\mathrm{e} \tag{4.26b}$$

and the SNR is now given by

$$\mathrm{SNR} = \frac{B_\mathrm{max} - B_\mathrm{min}}{B_\mathrm{min}} \tag{4.27}$$

Evaluation of equation (4.27) shows that the SNR has a maximum value when $k = 1$ and $n_\mathrm{e} = 0$ and falls off quickly as k increases or decreases. These fringes have intrinsically poor contrast so that correlation fringes are not observed with a two-speckled-beam system using addition. Since time-averaged fringes are observed by the addition process, it is seen that conventional two-speckled-beams systems cannot be used to observe vibration. These arguments apply for the usual circular viewing lens aperture. Ek and Biedermann (2) have shown that the use of a twin slit aperture enables the $4\sigma_\mathrm{s}{}^2$ term to be eliminated. Slettemoen (10) has extended this approach and developed a masked slit aperture which enables time-averaged fringes to be observed using a speckled reference wavefront.

4.4.5 Smooth reference beam interferometer: addition fringes

In the system, the monitor brightness is given by

$$B = q\langle I_1\rangle\left|1 + k^2\gamma^2 + 8k \cos^2 \tfrac{1}{2}\Delta\phi\right|^{\frac{1}{2}} + n_\mathrm{e} \tag{4.28}$$

If it is assumed that $8k \gg 1 + k^2\gamma^2$, the maximum brightness may be written as

$$B_{\max} \simeq q\langle I_1\rangle\sqrt{8k} \qquad (4.29)$$

and the SNR is now given by

$$\text{SNR} = \frac{\sqrt{8k}}{(1 + k^2\gamma^2)^{\frac{1}{3}} + n_e/q\langle I_1\rangle} \qquad (4.30)$$

The SNR can be shown to have a maximum value when $k\gamma = 1$, and since

$$k = \frac{\langle I_2\rangle}{\langle I_1\rangle}\gamma_2 = \frac{\langle I_2\rangle}{\sigma_2} \qquad (4.31a)$$

we have

$$\langle I_1\rangle = \sigma_2 \qquad (4.31b)$$

Thus, the mean value of the intensity of the speckled beam should equal the standard deviation or the rms noise in the reference beam. (This result was originally derived by Slettemoen. (6))

It can be seen from equation (4.31) that the magnitude of the optimum SNR increases as k increases (i.e. as γ decreases). Thus, the optical noise in the reference beam should be minimized to give the best fringes.

To minimize the electronic noise term, $n_e/q\langle I_1\rangle$ in equation (4.30) it is clear that the value of $\langle I_1\rangle$ should be maximized. Thus, ρ in equation (4.23) should be unity.

In the smooth reference beam speckle interferometer where addition is used to observe the fringes, the following conditions should be satisfied:

    (i) The reference beam should be as noise free as possible (clean optical components and a pinhole filter should be used).

    (ii) The mean intensity of the speckled (object) beam should equal the rms noise of the reference beam.

    (iii) The sum of the intensities should be such that the peak intensities are approximately at the saturation level of the camera.

4.5 Optical design of Electronic Speckle Pattern Correlation Interferometers

The principal kinds of measurement which are made using electronic speckle pattern correlation interferometers are:

    (i) in-plane displacements (Section 3.7.1);

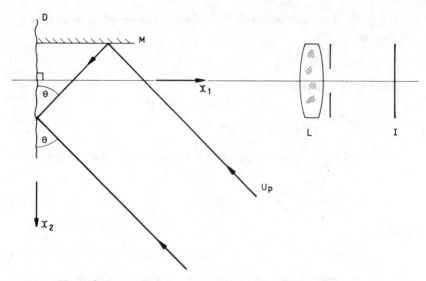

Fig. 4.6 A practical arrangement of an in-plane displacement sensitive speckle pattern correlation interferometer

(ii) out-of-plane displacements (Section 3.7.3);

(iii) shape measurement (Chapter 5).

The first of these requires two plane wavefronts to be incident at equal and opposite angles to the object surface: optical arrangements for doing this are discussed in 4.5.1. The other two require an in-line smooth reference beam, and such systems are discussed in 4.5.2.

4.5.1 *In-plane displacement sensitive interferometers*

Two versions of the basic plane-strain sensitive arrangement (Figure 3.17) are shown in Figures 4.6 and 4.7. In the first of these the two object illumination wavefronts U_o' and U_o'' are obtained by allowing half the incident plane wavefront U_p to illuminate the object D directly. The remainder of the wavefront is reflected onto the surface by the mirror M. When it is inconvenient to place a mirror in the direct vicinity of the object the arrangement shown in Figure 4.7 may be used. Here the plane wavefront U_p is split into two components U_o' and U_o'' by the right angled front reflecting prism, P. These are then reflected onto the object by the mirrors M_1 and M_2. Light scattered by the object reflects from the mirror M_3 onto the viewing lens L. In either of these cases care should be taken to obtain uniform object illumination. Any

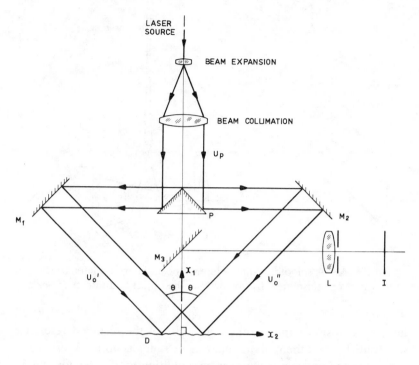

Fig. 4.7 An alternative arrangement of in-plane displacement sensitive speckle pattern correlation interferometer

significant fall-off in intensity at the edge of the illuminated area due to the Gaussian distribution of the laser intensity profile (Section 6.3.2) reduces the video signal to noise ratio in that region and will cause a reduction in fringe visibility. This effect may be avoided by making the initial expansion of the laser beam sufficient to ensure that only the central portion is used to illuminate the object. A 50 mm diameter F 4 collimating lens used in combination with a ×40 microscope objective expansion lens is commonly used. (See Section 6.3.2 for a detailed discussion of laser beam expansion.)

4.5.2 *Smooth reference beam systems*

There are two basic ways of combining the object and reference wavefronts. These are shown in Figures 4.8 and 4.9 respectively.

In Figure 4.9, the unexpanded input reference beam is pre-expanded by the negative lens L_1 and then focussed down through a 25 μm pinhole by the microscope objective lens (or a similar precision element), L_2.

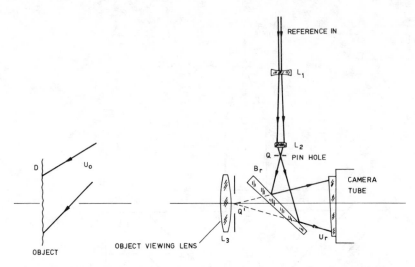

Fig. 4.8 A smooth, in-line reference beam interferometer which
uses a beamsplitter, B_1, to combine the object and reference beams

The pre-expansion improves the uniformity of the resultant reference
wavefront, U_r, and the pinhole removes extraneous optical noise. (This
is particularly important for use in the addition mode, Section 4.4.5.)
U_r is reflected onto the face plate of the camera by the small angle
wedge ($\simeq 1°$) beam splitter B_r. Ideally, the latter is anti-reflection coated
on the rear face in order to suppress secondary reflections. Note that a
similar wedge is cemented to the face of the camera tube: this prevents
the formation of multiple internal reflection fringes at the face plate.
The point of divergence of the reference wavefront, Q, is made conjugate

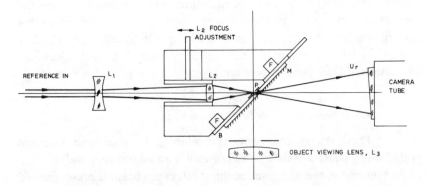

Fig. 4.9 A smooth, in-line reference beam interferometer based on
the hole-in-mirror principle

with the centre of the viewing lens, Q', i.e. the departure from conjugacy Δl (equation 4.21) is nominally zero. As the focal length of the viewing lens decreases, so does the image-plane distance l_o (Figure 4.4) until a point is reached beyond which it becomes impractical to mount the beamsplitter in the space available between the viewing lens and photo-cathode. The minimum practical focal length is $\simeq 40$ mm. Light scattered from the object is collected by the viewing lens L_3 and an image is formed in the plane I where it interferes with U_r. This system suffers from the disadvantage that dust particles which tend to collect upon the beamsplitter, together with any small blemishes due to imperfect clean-ing, act as light scattering centres. These cause reference beam noise and hence degrade the quality of addition fringes (Section 4.4.5) and two-wavelength-shape-difference fringes (Sections 5.4–5.6.1). In the latter case the change of scattering characteristics with wavelength causes both subtraction and addition fringes to be affected.

An alternative arrangement which considerably reduces the reference beam noise is shown in Figure 4.9. The unexpanded reference beam is pre-expanded by L_1 and focussed down through a 25 μm pinhole, P, mounted behind a hole of approximately 300 μm diameter drilled at 45° through a front reflecting mirror of thickness 0.5 mm. Note that the pinhole is made of sufficiently thin foil for it to be operated effectively at 45°. This pinhole mirror combination is mounted on a steel backing plate B which is held in position by small magnets F. The reference beam is aligned by translating the lens so that its focal point coincides with the pinhole plane, and sliding the mirror–pinhole combination into position on the magnets.

The small hole in the mirror does not degrade the image quality unless a significant fraction of the light transmitted by the viewing lens L_3 is incident on the hole. This will occur when the focal length of the viewing lens is 50 mm or less and the aperture is sufficiently small that the television camera at least partly resolves the speckle pattern (Section 4.3). Thus, this method of obtaining a smooth reference beam must be used with a viewing lens whose focal length is greater than 50 mm and if a large object is being viewed, it will have to be located at a considerable distance from the viewing lens.

The point of divergence of the reference beam and the centre of the viewing lens are clearly not conjugate in this arrangement but the departure from conjugacy is within the limits specified by equation (4.22) when the focal length of the viewing lens is greater than 60 mm.

The object beam illumination is usually obtained by expanding the laser beam to a suitable size so that it illuminates the object (Section

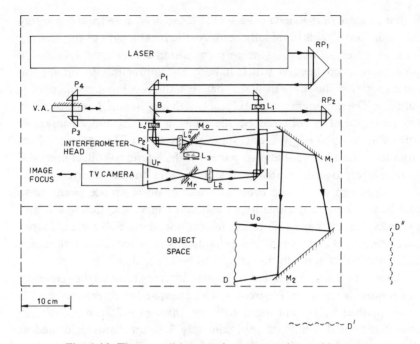

Fig. 4.10 The overall layout of an electronic speckle pattern interferometer with on-axis object-beam illumination. (VA is a variable attenuator which enables the reference to object beam ratio to be set at the optimum value, see Section 4.4)

6.3.2). In the complete interferometer layout shown in Figure 4.10 the point of divergence of the object beam is in-line and nearly conjugate with the centre of the viewing lens aperture. This is a requirement of some two-wavelength shape-measurement interferometers (see Section 5.5) and also maximizes the image intensity when the object surface is not diffuse since the object is then viewed in the specular direction.

In this arrangement in which the object beam has been folded in order to make the system as compact as possible, the unexpanded laser beam is directed into the optics via the retro prism RP_1 and the right angled prism P_1, and is split at a wedged beamsplitter B. Here, approximately 95 % of the light is transmitted and expanded by the L_o', L_o'' lens combination to form the object illumination wavefront U_o. This passes through the small hole in the mirror M_o and is reflected by M_1 and M_2 onto the object D. The light is then scattered by the object surface, and some of this scattered light is reflected back by M_2, M_1 and M_o onto the viewing lens L_3 which forms an image on the television camera via the mirror M_R, M_o and M_R are both identical in design to the mirror shown in Figure 4.9. The light which is reflected at B forms the reference beam. It is reflected by RP_2 onto the prisms P_3 and P_4 whose positions may be adjusted in order to match the coherence length of the object and reference beams arriving at the camera. It is then expanded by L_1 and focussed down through the small hole in the mirror M_r by L_2 onto the camera face where it interferes with the object beam. This interferometer does not represent the simplest configuration but illustrates some design compromises that have to be made in order to minimize the size of the system. Viewing distances of up to 1 metre may be accommodated within the base plate dimensions of approximately $500\,mm \times 600\,mm$. Furthermore the mirror M_2 may be adjusted to enable objects lying in the plane D' and D" and the plane of the figure to be observed. The interferometer head is suitable for double as well as single wavelength applications.

4.6 Video system considerations in Electronic Speckle Pattern Correlation Interferometers

The video system comprises the television camera, video store (if used), signal processing unit and display monitor. The aim of the system design is to obtain maximum visibility fringes. Electronic noise and, in particular, high frequency noise should be minimized. Since the cost of a laser increases approximately in proportion to the output power, it is important to optimize the sensitivity of the video system.

The values of I_{sat} for various camera tubes which are commercially available are given in Table 4.1. Of these, the Isocon tube clearly has the highest sensitivity; however, it is very much more expensive than the others but may be used for the inspection of large objects when the laser power is limited. The Chalnicon, Newvicon and Silicon tubes are all found to give satisfactory ESPI fringes and can be used to inspect diffusely scattering objects of up to about 100 mm in diameter using a 5 mW HeNe laser. (See Section 6.3.3 for a detailed discussion of the light scattering properties of different surface finishes.)

The most noise sensitive part of the signal processing is at the first stage of video signal amplification which is carred out at the camera head amplifier so that this section of the system must be carefully designed. It has been found to be useful to DC-couple the output signal into a high impedance device such as a FET which is connected directly to the face plate.

Table 4.1 *Saturation light intensities and relative sensitivities of TV tubes*

Tube type	I_{sat} (μw cm^{-2})	Relative sensitivity
EMI 9677C (Standard Vidicon)	0.53	1
EMI 9877M (Standard Vidicon)	0.59	0.9
RCA 4542 (Long-lag tube)	0.08	6.6
Thomson 7H9828 (Silicon target)	0.21	2.5
RCA 4532 (Silicon target)	0.24	2.2
Mullard XQ 1440 (Newvicon)	0.18	2.9
Isocon	0.005	$\approx$100

(Sensitivities measured at $\lambda = 633$ nm using a HeNe laser)

It was seen in Section 4.3 that the size of the correlation signal, and hence the SNR of the fringes is a function of the spatial resolution of the video system, and the SNR is maximized if the spatial resolution is maximized. The vertical resolution is determined by the number of scan lines, so that when a standard video system is used this is fixed. The frequency response of the system should be such that the horizontal spatial resolution is as good as the vertical resolution.

When correlation fringes are observed using subtraction it is necessary to lock the timing pulses of the live camera picture to those of the stored picture. If this is not done to a high degree of precision a loss of correlation and, hence, fringe contrast will occur. When a video tape recorder or video disc is used, the timing pulses for the stored picture are generated mechanically and as a result tend to fluctuate slightly in time. These fluctuations cannot always be followed accurately by the camera locking circuitry and can result in reduced fringe visibility. A solid state video store requires high speed A to D/D to A conversion for storage and display. The timing pulses in this case can be derived from a single source and good correlation between live and subtracted signals is thereby maintained. As the cost of solid state stores decreases and their availability increases, they are being used increasingly in preference to analogue recorders or discs.

Another important aspect of the video processing is the design of the high-pass filter. The most successful form of filter has been based on an R–C network designed to have a 3 dB break point at an upper frequency corresponding to the maximum speckle frequency ($\sim$20 μm corresponding to 4 MHz), followed by a roll off at 6 dB per octave. The low frequency cut off is typically set at 500 KHz. Subtle changes in filter characteristic have been found to have quite significant effects upon the appearance of the speckle pattern and the correlation fringes. For example, the basic R–C network may be modified so that it differentiates the video signal at each edge. This gives an apparently finer speckle pattern display and a subjective improvement in fringe quality. The complexity of the transfer function between camera signal and monitor output makes it difficult to quantify this aspect of the electronic design.

Correlation fringes can be observed on any television monitor. However, to give reasonable visibility, the monitor should have a good contrast and brightness range, and should also maintain its focus over the whole screen. If accurate measurements are to be made, there should be a minimum amount of distortion.

4.7 Dynamic displacement measurements

The three basic techniques which are used in holographic interferometry to measure dynamic displacements can also be applied to ESPI as follows:

(i) time-averaged vibration ESPI;
(ii) stroboscopic ESPI;
(iii) pulsed laser ESPI.

An additional technique developed by Hogmoen and Løkberg (4) which is an extension of time-averaged stroboscopic ESPI enables the amplitude of vibration to be measured at very high sensitivity (<1 nm). This relies on the phase modulation of the reference beam.

These techniques are discussed in the following sections.

4.7.1 *Time-averaged vibration ESPI*

Fringes which represent the variation in the amplitude of the out-of-plane vibration of a surface are observed with a smooth in-line reference beam electronic speckle system (see Sections 4.2.2 and 4.4.5). The shape and amplitude of the resonant modes can be found in real-time by varying the frequency and amplitude of the object excitation; the technique thus provides a very useful method for routine vibration analysis (see also Section 7.5).

The disadvantages of the method are that:

(*a*) the fringe visibility decreases as the amplitude of vibration increases (they have a J_0^2 envelope for a sinusoidal vibration); and

(*b*) the fringes represent only the amplitude and not the phase of the vibration. (See equation (4.11) and Figure 4.3.)

4.7.2 *Stroboscopic ESPI*

The visibility of time-averaged vibration fringes is considerably improved if the laser beam is amplitude modulated at source; the light is chopped so that it produces pulses at two points t' and t'' separated by a time Δt during the vibration cycle (see Figure 4.11). Fringes having a $\cos^2$ intensity distribution defining the change in displacement that occurs during Δt are then observed as a result of signal addition (Section 4.2.2). This displacement depends on the phase of the pulses with respect

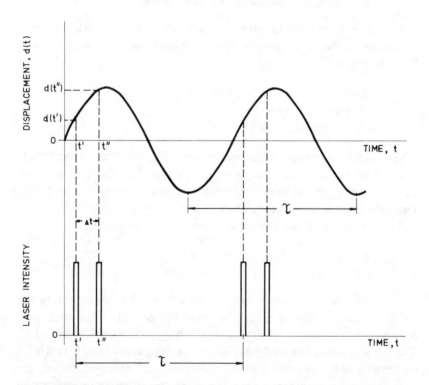

Fig. 4.11 The form of laser beam intensity modulation for the observation of stroboscopic $\cos^2$ addition fringes

to the object vibration phase, so that if the phase and separation of the two pulses are varied, the phase as well as the amplitude of the object vibration can be determined.

This procedure is complicated and time consuming; a quicker, more flexible technique due to Løkberg and Høgmoen (3) is based on modulating the optical path length of the reference beam (for example by oscillating one of the mirrors in the reference beam path) at the same frequency as the object vibration.

The fringe information term of the time-averaged fringe pattern has the general form

$$S = 8\langle I_r \rangle \langle I_s \rangle J_0^2 \left\{ [a_s^2 + a_r^2 - 2a_s a_r \cos(\phi_s - \phi_r)]^{\frac{1}{2}} \frac{4\pi}{\lambda} \right\} \qquad (4.32)$$

where a_s and ϕ_s are the amplitude and phase of the object vibration, and are functions of object coordinate, and a_r, ϕ_r are the amplitude and phase of the reference beam path modulation. In general, equation (4.32) represents a complicated fringe pattern. However, for regions where $a_r = a_s$, and $\phi_r = \phi_s$ the argument of the Bessel function becomes zero; hence the zero order maximum brightness fringe coincides with those parts of the object which are vibrating in phase with the reference beam. When $a_r \neq a_s$ the Bessel function will still have a minimum value when $\phi_r = \phi_s$ since $2a_s a_r \cos(\phi_0 - \phi_s)$ will have a maximum value. In general, therefore, the phase of the vibration can be mapped out by setting a_r at a given value and noting the region of maximum brightness as ϕ_r is varied. a_r may be adjusted to maximize the brightness of the zero order fringe.

4.7.3 *High sensitivity measurements of vibration amplitude* (4)

It is possible to extend the last method to the measurement of sub-wavelength vibration amplitudes. To do this, ϕ_r is varied continuously by modulating the reference beam path at a frequency $f + \Delta f$, where f is the frequency of the object vibration, and Δf is less than the television frame rate. Equation (4.32) can be written as

$$S = 8\langle I_r \rangle \langle I_s \rangle J_0^2 \left(\frac{4\pi}{\lambda} X \right) \qquad (4.33)$$

where

$$X = \{ a_s^2 + a_r^2 - 2a_s a_s \cos[\phi_s - \phi_r(t)] \}^{\frac{1}{2}}$$

At those points in the object where $a_s = 0$, the value of X is constant, but when $a_s \neq 0$, the value of X varies in time; this will be observed as

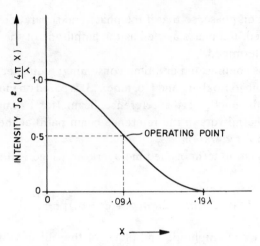

Fig. 4.12 The maximum sensitivity operating point for small amplitude displacement measurement ($\lambda = 633$ nm). The amplitude of the reference beam phase modulation, a_r, is set such that intensity variations occur over the linear region of $J_0^2(X)$

a flickering at frequency Δf of the intensity on the television monitor. X has a mean value of $(a_r^2 + a_s^2)^{\frac{1}{2}}$, which is approximately equal to a_r when $a_s \ll a_r$, and fluctuates between minimum and maximum values of $(a_r - a_s)$ and $(a_r + a_s)$. It can be seen from Figure 4.12 that the amplitude of the intensity fluctuation is maximized if the mean value of X is such that the mean value of the intensity corresponds to the $J_0^2(X) = 0.5$ position; when $a_s \ll a_r$, this condition is satisfied when $a_r = 0.9 \lambda$.

The presence or absence of flicker on the monitor will indicate whether the vibration is above or below a threshold level; this level must be determined empirically. While the human eye should theoretically be able to detect a 1 % change in intensity at 8–12 Hz, it has been found in practice (4) that the change in intensity must be ~6%, corresponding to an amplitude of ~2 nm to be detected by eye.

This technique may be quantified by using a photodetector placed in contact with the television monitor screen to measure the intensity variations. The area of such a detector should be large enough to average the intensity over a few speckles and will give an output which is a complicated function of the television camera transfer function (7). This means that the system must be calibrated using a modulating mirror of known response with the object at rest. When this is done vibration amplitudes of the order of 10^{-2} nm have been measured (4).

4.7.4 *ESPI using a pulsed laser*

A dual pulsed laser in conjunction with a smooth reference beam interferometer (8, 9) can be used to give out-of-plane ESPI fringes which are equivalent to the fringes obtained with the dual pulsed holographic system (Section 2.6.3). The television camera scan is switched off and the speckle patterns derived from the two pulses are added on the television face plate. The camera scan is then switched on.

Since the speckle pattern then exists on the camera for only a small number of scan frames, the output needs to be recorded in order to obtain a permanent record. This can be done using a video tape recorder or a single-frame store which has been set to record at a set interval after camera scan switch-on corresponding to the frame of optimum fringe contrast. An experiment using this system is described in Section 7.5.2.

In-plane measurements use two speckled beams and therefore require the use of the subtraction process. Thus, ESPI measurements of in-plane displacement using a pulsed system are made by storing the speckle pattern for the static object and subtracting the speckle pattern for the displaced object. This technique is useful for measuring in-plane displacements of large areas and for inspecting objects which are moving rapidly, for example objects which are rotating at high speeds.

4.8 Limitations of Electronic Speckle Pattern Interferometric measurements

We now look at the main factors that limit the range of measurements that can be made using the various electronic speckle pattern correlation interferometers.

4.8.1 *Measurement sensitivity*

ESPI can be used to give fringes which represent lines of either in-plane or out-of-plane displacement. The fringe sensitivity for an in-plane interferometer is given by $\lambda/(2 \sin \theta)$ where λ is the wavelength of the light used and θ is the angle of incidence of the illuminating beams. Out-of-plane interferometers may give fringes representing constant displacements at intervals of the order of $\frac{1}{2}\lambda$, or may be desensitized up to about 100 μm. An interferometer giving fringes representing loci of constant out-of-plane displacement gradients can also be set up (Section 3.7.2).

It is not possible to detect less than one fringe accurately with a conventional ESPI system so that this represents the minimum sensitivity of the system. The technique discussed in 4.7.3 enables very small amplitude (≈ 0.01 nm) vibrations to be measured using time-averaged ESPI.

It was seen in 4.3.4 that the fringe spacing on the television camera must be less than 1/120th of the screen width; this limits the displacement gradient and also the total displacement which can be observed. The visibility of the fringes obtained using time-averaged ESPI to observe out-of-plane vibrations falls off rapidly with increasing vibration amplitude unless stroboscopic illumination is used; the maximum number of fringes observed with a continuous laser source in a time-averaged system is at present 14 (10).

It should be noted that only out-of-plane sensitive time-averaged fringes can be observed.

4.8.2 *Object size limitations*

The maximum area which can be inspected in one view is limited by the laser power available and the camera sensitivity (see Section 4.4). Out-of-plane displacement fringes have recently been observed on an area of 0.75 m in diameter at Loughborough using a camera with a Newvicon tube and a 15 mW HeNe laser. There is no reason why a larger area should not be inspected if sufficient laser power is available; the mechanical stability of the system and the coherence of the laser (see Section 6.3) also limit the performance of the system.

When in-plane measurements are being made, the illuminating wavefronts must be plane if fringes of uniform sensitivity are to be obtained. Unless a large collimating lens is used for large areas, the fringe pattern interpretation will be rather difficult. When the surface of the object being inspected is not flat, it can be shown that the fringes may have sensitivity to out-of-plane movements.

In order to observe fringes on a small area, a relatively large deforming force must be applied, and this is likely to give rise to rigid body translations. Rigid body translations cause holographic fringes to localize away from the object surface, but in speckle correlation interferometry, they cause speckle decorrelation and hence a reduction in fringe visibility. The same applies to displacements to which the interferometer does not have fringe sensitivity. Decorrelation effects are discussed in Appendix F, and it is seen that decorrelation due to in-plane translations and rotations increases as the magnification of the viewing system increases.

Thus, decorrelation of the speckle, causing a reduction in fringe visibility, is likely when ESPI is used at high magnification. Fringes have been observed on an area of $0.05 \, \text{mm}^2$ in an out-of-plane ESPI system (11) and on an area of $0.25 \, \text{mm}^2$ in an in-plane system by one of the authors (C.W.).

4.8.3 Depth of field

Because the ESPI system must use a high F-number (small lens aperture) viewing system, the depth of field of the system is high, and so this is generally not a restriction in ESPI.

4.9 Summary

Electronic Speckle Pattern Correlation Interferometry (ESPI) is a technique which enables real-time speckle pattern correlation fringes to be observed instantaneously. A television camera and video signal processor replace the photographic methods described in Chapter 3. The electronic processing may be used in conjunction with any type of speckle pattern correlation interferometer and the fringes are formed by video signal subtraction or addition. In the latter case the interferometer should have smooth reference wavefronts. This chapter has described the basic principles of these methods and shows how the best performance may be obtained from the various systems. Perhaps the most important aspect of the technique is the ease with which the fringes may be obtained in comparison with the photographic methods. This enables speckle pattern correlation interferometry to be extended to the solution of quite complex problems in displacement (Chapter 7) and shape measurement (Chapter 5).

4.10 Suggestions for further reading

The following references provide a useful reading in addition to the material presented in this chapter:

System design and optimization: References (12)–(14).

In-plane strain measurement and associated speckle pattern decorrelation phenomena: References (15), (16).

Two-wavelength applications in shape measurement and associated speckle pattern decorrelation phenomena (see also Chapter 5): References (17)–(19)

Review articles: Reference (20).

5

Holographic and speckle pattern interferometric techniques for shape measurement

5.1 Introduction

Conventional shape measuring instruments use mechanical probes and give either point-by-point or line-scan information about shape (1). An optical method of measuring shape has the advantage of being non-contacting and can also give a field view of the surface under investigation. Thus, there has been considerable effort directed towards the development of optical shape-measurement techniques.

Holographic methods of measuring surface shape are based on a two-wavelength technique first reported by Hildebrand and Haines (2). The two wavelengths can be produced by using two laser lines of different frequencies, or alternatively by altering the refractive index of the medium surrounding the object. The fringes represent the intersection of the object surface with a set of surfaces which in general are hyperboloids, but may be a set of equispaced planes in which case the fringes represent true depth contours. A new hologram must be made each time a new component is inspected.

ESPI can be used to compare the shape of test components with a master wavefront. The fringes obtained represent the difference in depth along the viewing direction between the master wavefront and the test component. The master wavefront may be produced by conventional optical components (i.e. flat, spherical or cylindrical) or may be generated holographically using a master component. The system enables components to be inspected in rapid succession.

When two beams of light which interfere to form a fringe pattern are projected onto the surface of an object, the form of fringes observed on the object surface depends on the shape of the surfaces (3). Abramson has applied his sandwich holography technique (4) (see also Section 6.6.2) so that the projected fringes can be rotated to give fringe surfaces which are parallel to any required plane. This is done by rotating the sandwich hologram and is discussed in Section 5.9.1.

Another technique first described by Brooks and Heflinger (5) which uses projected fringes to compare the shape of a test component with a master component is discussed in Section 5.9.2. Although this is neither a holographic nor a speckle technique it is discussed here because it is analogous and complementary to the ESPI shape-measurement system.

5.2 Two-wavelength holographic contouring

The use of two-wavelength holography for measuring the surface .shape of objects with optically rough surfaces was first reported by Hildebrand and Haines (2). Subsequent work has modified the technique to simplify the interpretation of the fringes (6–14).

The technique requires that a holographic recording be made of an object at two wavelengths, λ_1 and λ_2, on a single hologram plate. The developed plate is then illuminated by the reference beam at only one wavelength λ_1, so that the original object beam at λ_1 is reconstructed together with a distorted reconstruction of the object beam at λ_2. When the hologram is viewed, these two wavefronts may interfere to give fringes which are related to the shape of the object surface.

An alternative method of obtaining the two wavefronts is to make a hologram of the object at λ_2 only, and then to illuminate both the hologram and the object at λ_1, giving the undistorted object beam at λ_1, and the distorted reconstruction of the object beam at λ_2.

The fringes observed represent the intersection of the object surface with a set of surfaces which are defined by the illumination and viewing geometries as well as by the difference in wavelength $(\lambda_1 - \lambda_2)$, so that the interpretation of the fringes can be very difficult. In some circumstances, the fringes may be made to represent the intersection of the object surface with a set of equispaced planes, in which case they represent true depth contours; this enables absolute shape measurements to be made. However, objects which have large curvature, and hence considerable depth give large numbers of fringes when measured at high sensitivity, and again the analysis can be difficult.

A new hologram must be made for each component inspected. This is quite laborious using normal holographic emulsion. Friesem *et al.* (14) have performed contouring measurements using photoplastic recording materials which can be developed in a short time (~3 s) *in situ* by heating. This means that rapid inspection of many components can be performed.

At present, the recording area achievable with photoplastic materials is considerably less than that of conventional photographic plates.

The two wavelengths may be obtained using a multifrequency laser such as an Argon laser, giving contours in the sensitivity range from 2–30 μm. Tunable dye lasers enable contours of up to several millimetres to be obtained since they can produce very closely spaced frequencies (see Section 6.3.5 and Table 6.2).

The change in wavelength may also be achieved by altering the refractive index of the medium surrounding the object. This is done by placing the object in a glass tank into which fluids of varying refractive index are added. A change in wavelength is associated with the change in refractive index (see Section 1.5.5) and sensitivities from 1.5 μm to several millimetres may be obtained.

5.2.1 *Theoretical and practical aspects of two-wavelength holographic contouring*

The object and reference beam amplitudes at a point on the hologram plate ($U_{o1,2}$, $U_{r1,2}$, at $\lambda_{1,2}$) are given by

$$U_{o1,2} = u_o \exp\left(\frac{2\pi i}{\lambda_{1,2}}l_o + \phi_s\right) \tag{5.1}$$

$$U_{r1,2} = u_r \exp\left(\frac{2\pi i}{\lambda_{1,2}}l_r\right) \tag{5.2}$$

where ($l_o - l_r$) represents the difference in path travelled by the object and reference beams, and ϕ_s represents the random phase associated with the rough surface. This is assumed to be the same for λ_1 and λ_2 (see Appendix F).

The intensity at that point on the hologram plate is then given by

$$I_{1,2} \propto U_o U_o^* + U_r U_r^* + U_o U_r^* + U_r U_o^*$$

where $U_{o1,2}$ and $U_{r1,2}$ are written as U_o, U_r for brevity.

When the developed plate is illuminated by the reference beam at λ_1, the reconstructed object beam is given from Section 1.7.3 by

$$U_{\text{reconst}} = U_{r_1} U_o U_r^*$$

The two wavefronts are then given by

$$U_1 = u_o u_r^2 \exp\left[2\pi i\left(\frac{l_o - l_r}{\lambda_1} + \frac{l_r}{\lambda_1}\right) + \phi_s\right] \tag{5.3}$$

$$U_2 = u_o u_r{}^2 \exp\left[2\pi i\left(\frac{l_o - l_r}{\lambda_2} + \frac{l_r}{\lambda_1}\right) + \phi_s\right] \tag{5.4}$$

Thus, the phase difference, $\Delta\phi$, between the two wavefronts is

$$\Delta\phi = 2\pi\left(\frac{1}{\lambda_1} - \frac{1}{\lambda_2}\right)(l_o - l_r) \tag{5.5}$$

If the two wavefronts are changed by altering the refractive index, equation (5.1) becomes

$$U_{o1,2} = u_o \exp\left[\frac{2\pi i}{\lambda}\left(\frac{l_o{}'}{n_{1,2}} + l_o{}''\right) + \phi_s\right] \tag{5.6}$$

where $l_o{}'$ is the distance travelled by the object beam in the cell, and $l_o{}'' = l_o - l_o{}'$. The reconstructed waves are now given by:

$$U_1 = u_o u_r{}^2 \exp\left[\frac{2\pi i}{\lambda}\left(\frac{l_o{}'}{n_1} + l_o{}'' - l_r\right) + \phi_s\right] \tag{5.7}$$

$$U_2 = u_o u_r{}^2 \exp\left[\frac{2\pi i}{\lambda}\left(\frac{l_o{}'}{n_2} + l_o{}'' - l_r\right) + \phi_s\right] \tag{5.8}$$

and hence the phase difference is

$$\Delta\phi = \frac{2\pi}{\lambda}\left(\frac{1}{n_1} - \frac{1}{n_2}\right)l_o{}' \tag{5.9}$$

When the object is viewed through the hologram, two images are obtained corresponding to the two wavefronts given by equations (5.3) and (5.4) or equations (5.7) and (5.8). Under suitable viewing conditions, these two wavefronts will interfere to give fringes at intervals of

$$\Delta(l_o - l_r) = \frac{\lambda_1 \lambda_2}{(\lambda_1 - \lambda_2)} \tag{5.10}$$

or

$$\Delta l'_o = \frac{n_1 n_2}{(n_1 - n_2)} \tag{5.11}$$

Clearly in either case, the spacing of the fringes depends on the direction and form of the object illumination beam, and also on the direction in which the hologram is viewed as well as on the object shape. This indicates the difficulty associated with the interpretation of the fringe pattern.

When a two-frequency laser is used to give two wavelengths, equivalent points in the two images may be displaced since the wavefront reconstructed by the reference beam at λ_1 from the λ_2 hologram is distorted. The viewing system must have a resolution diameter in the image plane which is greater than this displacement if the two images are to interfere. This can give rise to a very speckly fringe pattern due to the small lens aperture which may be necessary. This effect is best minimized by having the object as close to the hologram as possible. (This problem does not arise when the dual refractive index method is used.)

The system described by Varner (12) simplifies the interpretation and minimizes the image displacement. A telecentric viewing system is used to form the image (see Figure 5.1). In such a system, two lenses are placed so that their focal planes coincide, and the aperture is located in this plane. This arrangement ensures that only those rays in and around the axial direction contribute to the image. The fringes can then be seen only over a very small range of angles and the fringe sensitivity becomes well defined. This arrangement also ensures that the object is effectively very close to the hologram plane, so that the relative shift in the two images is minimized.

It can be shown that if plane waves are used for both object and reference beams, the contours represent the intersection of the object surface with equispaced planes of spacing

$$\Omega = \frac{\lambda_1\lambda_2}{(\lambda_1-\lambda_2)(1+\cos\theta_o)} \tag{5.12}$$

where θ_o is the angle of incidence of the object illumination beam. If the object and reference beams are arranged such that $\theta_o = -\theta_r$ (see Figure 5.1) the fringe planes are parallel to the hologram plane. (θ_r is

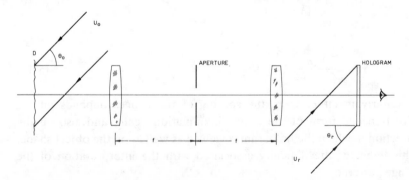

Fig. 5.1 The telecentric viewing system for the observation of holographic contour fringes; see Varner (12)

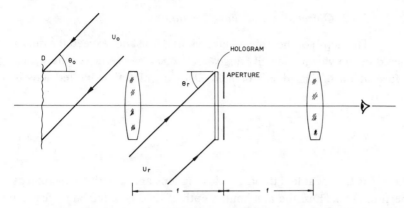

Fig. 5.2 An adaptation of the telecentric viewing system shown in
Figure 5.1 for use with either a large object or small hologram
recording area. (See Zelenka and Varner (10)). Note that the
hologram plate is now located near the aperture

the angle of incidence of the holographic reference beam to the hologram
plate).

If the object being viewed is large, or if the holographic recording
medium covers only a small area, it may not be possible to locate the
hologram plate near the image. In this case, the arrangement shown in
Figure 5.2 also due to Zelenka and Varner (10), may be used. The plate
is now located near the aperture. This system again has a well defined
fringe sensitivity. The fringes will represent the intersection of the object
with planes parallel to the hologram plane when the object is illuminated
normally using a beamsplitter. A small aperture may, however, be
necessary to overcome the relative displacement of the two images when
two frequencies are used, giving relatively large speckles. This image
shift does not occur when the dual refractive index method is used, so
that this arrangement is particularly suitable since it is more efficient in
light usage than the arrangement shown in Figure 5.1.

5.3 Limitations of holographic contouring

5.3.1 *Sensitivity*

The sensitivity of the contour fringes can be varied from 1 μm
up to several millimetres using either of the two-wavelength techniques
with a suitable choice of either frequency interval, or refractive index
difference.

5.3.2 Object size and shape limitations

The depth of the object being viewed must not exceed the depth of field of the viewing lens if fringes are to be observed over the whole surface at once. The depth of field D of a diffraction limited lens is given by

$$D = \lambda\left(\frac{f}{a}\right)^2 \qquad (5.13)$$

where f is the focal length of the viewing lens and a is the viewing lens aperture (15). Thus, the maximum depth of field is obtained by using a minimum lens aperture.

Since the spacing of the fringes is proportional to the slope of the object surface, this slope must be less than the value which gives fringes equal to the speckle size or the fringes will vanish (see Appendix F); since the speckle size is proportional to f/a, it is seen that the maximum value of slope which can be observed is obtained when a maximum lens aperture is used.

Since these two requirements are contradictory, it will be appreciated that it is not always possible to inspect a given object in one view.

To have a well-defined fringe sensitivity, either the object must be located near the hologram plate and the hologram plate must be roughly the same size as the object, or else a lens must be used to form an image of the object on or near the hologram plate; this lens will have to have a large aperture if the light loss is not to be too great. To obtain fringes which represent the intersection of the object surface with a set of planes, plane wavefronts must be used for both object and reference beams. Thus, the inspection of a large object requires large collimating and imaging lenses, and/or a large hologram plate if the fringe interpretation is not to become very difficult.

5.3.3 Surface finish limitations

To form a holographic image of reasonably uniform intensity, the object surface must be diffusely scattering (see Section 6.3.3).

When high sensitivity contours are observed, speckle decorrelation is likely to occur due to the surface roughness (see Appendix F). Some decorrelation is also likely to occur because viewing is not in the specular direction; the greater the speckle size, the greater this decorrelation (see equation F.28).

5.4 Shape measurement using ESPI: basic principles

Electronic speckle pattern interferometry can be used to compare the shape of components having optically rough surfaces with master optical wavefronts (16, 17). The fringe pattern obtained gives the difference in depth along the viewing direction between the surface and the wavefront. A master wavefront which is flat, spherical or cylindrical may be produced using conventional optical components. Alternatively master wavefronts of complex shape can be reconstructed from a hologram made using a master component, or, in principle at least, from a computer generated hologram.

The test surface is ideally illuminated simultaneously by master wavefronts at wavelengths λ_1 and λ_2, and viewed in an electronic speckle pattern interferometer with a smooth in-line reference wavefront. Live correlation fringes are thereby obtained between the two speckle images as a result of addition (Section 4.2.2). The sensitivity of the fringes obtained (the contour interval) is determined by the wavelength difference $(\lambda_1 - \lambda_2)$ (see equation 5.25).

A pair of master wavefronts at λ_1 and λ_2 can be used to inspect a series of components of nominally identical shape by placing the component at the correct orientation in the illuminating wavefront. Live addition fringes may be obtained as described above. If this is not possible, the illuminations are performed sequentially, and the fringes are obtained by recording one speckle image and subtracting the second after the wavelength change (Section 4.2.1).

5.4.1 *The theoretical basis of ESPI shape measurement*

ESPI shape difference fringes are obtained by illuminating the test component with a master wavefront at two wavelengths, λ_1 and λ_2, either simultaneously or sequentially. An image of the object illuminated in this manner is formed on the face plate of a television camera and a smooth reference beam satisfying the conjugacy condition (Section 4.3.2) is added. When the object is illuminated simultaneously at the two wavelengths, addition correlation fringes are obtained. When the object is illuminated first at λ_1 and then at λ_2, the first image is recorded, and the second is subtracted from it to give fringes by subtraction.

The role of the master wavefront is fundamental to the ESPI shape measurement technique. It should first be noted that if the wavelength of illumination is changed in a conventional speckle interferometer – e.g. an out-of-plane sensitive system, as shown in Figure 4.10, the speckle

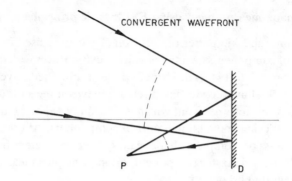

Fig. 5.3 The geometry of a spherical, convergent wavefront as used as a master illumination wavefront for a plane surface D

image at the second wavelength will be unrelated to that at the first, and no correlation fringes will be obtained. This can be seen by looking at the effect of altering λ in equation (4.15).

The basic requirement of a master wavefront is as follows. A master wavefront is one which converges to a point with a single phase when reflected by a surface of specularly reflecting finish whose shape matches the required master shape. A simple example is shown in Figure 5.3. The master surface is flat and when a converging spherical wavefront is reflected by a flat mirror, it converges to a point P. Thus the spherical wavefront acts as a master wavefront for a plane object.

An alternative arrangement which produces a flat master wavefront is shown in Figure 5.4. Here the plane surface is illuminated by an expanded collimated laser beam, and it can be seen that the reflected light will converge to the original point of divergence, P.

It will now be shown how the master wavefront can be used to compare test shapes with the master shape.

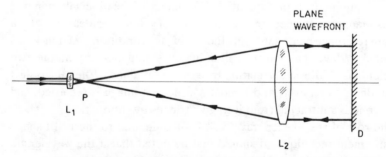

Fig. 5.4 The geometry of a plane wavefront as used as a master wavefront for a plane surface D

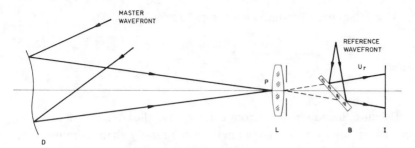

Fig. 5.5 The basic viewing arrangement for the inspection of a
component D using two-wavelength ESPI

Consider first what happens when a component whose shape is identical to the master shape but whose surface finish is optically rough is
illuminated by the master wavefronts. When it is illuminated at λ_1, an
image is formed on the face plate of a television camera by the viewing
lens L, whose centre is located at the point of convergence of the master
wavefront, see Figure 5.5. The smooth reference beam, U_r, is introduced
by the beam splitter, B.

The intensity of the light at a point in the image plane is given from
equation $(4.1a)$ as

$$\mathscr{I}_s = I_r + I_s + 2\sqrt{I_r I_s} \cos (\phi_o - \phi_r) \tag{5.14}$$

where $(\phi_o - \phi_r)$ represents the difference in phase between the object
and reference beams associated with the different optical paths travelled
by the two beams between the point where they are separated by a
beamsplitter to the point at which they are recombined at the face plate
of the television camera. ϕ_o may be written as

$$\phi_o = \phi_s + \phi_{q1} + \phi_{q2}$$

where ϕ_s is the random phase associated with the surface roughness,
ϕ_{q1} is the phase associated with the optical path of the object beam
between the beamsplitter and the centre of the viewing lens (this is the
path which the object beam would travel if the object were specularly
reflecting) and ϕ_{q2} is the optical path of the object beam from the centre
of the viewing lens to the image plane; these terms correspond to the
path length terms in equation (1.86). ϕ_r is the phase associated with
the optical path travelled by the reference beam between the point at
which the object and reference beams are split and the image plane.

As the reference beam is conjugate with the centre of the viewing
lens, $(\phi_{q2} - \phi_r)$ is constant across the image plane (see Section 4.3.2).

Thus, the intensity can be written as

$$\mathscr{I}_s = I_r + I_s + 2\sqrt{I_r I_s} \cos{(\phi_s - \psi_1)} \tag{5.15}$$

where

$$\psi_1 = \phi_{q1} + \phi_{q2} - \phi_r$$

Because the master wavefront converges to the centre of the viewing lens when reflected by a component of the master shape, the term ϕ_{q1} is constant for all points in the image, so that ψ_1 is also constant across the image.

Consider now what happens when the illuminating wavelength is changed. ψ_s can be assumed to remain constant (the conditions under which this assumption is valid are discussed in Appendix F.) The phase terms ψ'_{q1}, ψ'_{q2} and ψ'_r at λ_2 will have different values but the value of

$$\psi_2 = \psi'_{q1} + \psi'_{q2} - \psi'_r \tag{5.16}$$

will be constant for all points in the image plane.

If the two speckle patterns are correlated by subtraction, the resultant monitor brightness, B, is given by equation (4.3) as:

$$B = 4K\sqrt{\langle I_r \rangle \langle I_s \rangle} \sin^2{[\phi_s + \tfrac{1}{2}(\psi_1 - \psi_2)]} \sin^2{[\tfrac{1}{2}(\psi_1 - \psi_2)]} \tag{5.17}$$

When the correlation is obtained by addition the brightness B is given by equation (4.9) as

$$B = K[\sigma_r^2 + \sigma_s^2 + 2\langle I_r \rangle \langle I_s \rangle \cos^2{[\phi_s + \tfrac{1}{2}(\psi_1 + \psi_2)]} \sin^2{[\tfrac{1}{2}(\psi_1 - \psi_2)]}] \tag{5.18}$$

In both cases, the contrast of the speckle is constant across the image, since $(\psi_1 - \psi_2)$ is constant. Thus a component whose shape is that of the master shape gives a null fringe field.

Consider now what happens when a component whose surface shape is not the same as that of the master shape is illuminated by the master wavefronts.

In Figure 5.6, the test surface D is indicated by the heavy line, and the master surface by a dotted line. A point P in the object is illuminated by the master wavefront. The light in the image plane which has been scattered from P has travelled a distance Δl further than the light which would be scattered from an equivalent point in the master surface. (We can ignore the small lateral shift of the specularly reflected ray and the small change in angle, as the resultant shift in image position will be very much less than the speckle size). Δl is given by

$$\Delta l \simeq 2d \cos{\theta} \tag{5.19}$$

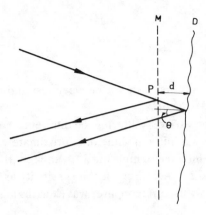

Fig. 5.6 The geometry for the formation of shape difference fringes between a master wavefront, M, and an inspection wavefront, D

where θ is the angle of incidence of the light and d is the normal separation of the test and master surfaces. When the illumination wavelength is λ_1, the intensity of the light in the image plane is given by

$$\mathcal{I}_1 = I_r + I_s + 2\sqrt{I_r I_s} \cos\left(\psi_1 + \frac{4\pi}{\lambda_1} d \cos\theta\right) \tag{5.20}$$

and at λ_2 the intensity is

$$\mathcal{I}_2 = I_r + I_s + 2\sqrt{I_r I_s} \cos\left(\psi_2 + \frac{4\pi}{\lambda_2} d \cos\theta\right) \tag{5.21}$$

where it is assumed that ϕ_s and I_s remain constant.

We see from (5.17) and (5.18) that a speckle correlation term of the form

$$S_{\text{corr}} \propto \cos^2 (\sin^2) \left[\tfrac{1}{2}(\psi_1 - \psi_2) + 2\pi \left(\frac{1}{\lambda_2} - \frac{1}{\lambda_1}\right) d \cos\theta \right] \tag{5.22}$$

is obtained, the $\cos^2$ term occurring when the correlation is obtained by the addition process and the $\sin^2$ term when subtraction is used. Thus, correlation fringes are obtained when the test surface shape differs from the master surface shape, the fringes occurring at intervals of $d = \Lambda$ given by

$$\Lambda = 1 \left/ \left[2\left(\frac{1}{\lambda_2} - \frac{1}{\lambda_1}\right) \cos\theta \right] \right. \tag{5.23}$$

which may be written as

$$\Lambda = \frac{\lambda_1 \lambda_2}{2(\lambda_1 - \lambda_2) \cos \theta} \tag{5.24}$$

Hence, the variation of d, the normal distance between test and master surfaces is mapped out at intervals of Λ.

The fringe sensitivity can be varied by altering the difference between the two wavelengths, and has a minimum value of approximately 2 μm. The upper limit could be as high as 10 mm, but it is unlikely that this technique would be employed for measurements of sensitivity as low as this. Values of Λ for commonly used Argon laser wavelengths are listed in Table 5.1.

5.4.2 *Fringe pattern interpretation*

It is seen from the fringe correlation term S_{corr}, equation (5.22), that the fringe pattern obtained in the shape measuring ESPI maps out the variation of the normal distance d between the test and master surfaces at intervals of

$$\Delta d = \frac{\lambda_1 \lambda_2}{2(\lambda_1 - \lambda_2) \cos \theta} \tag{5.25}$$

In order to determine the absolute difference in shape between test and master surfaces, several other parameters must also be measured.

Firstly, since θ is not necessarily constant across the surface, its value must be determined for the whole surface. This can be found by ray tracing or optically.

Second, since only the variation in d and not its absolute value is indicated by the fringe pattern, it is necessary to find the absolute value of the difference at one point. This can be done mechanically.

Table 5.1 Contour sensitivities for argon laser wavelength pairs

$\lambda_1/$(nm)	$\lambda_2/$(nm)	$\Lambda/$(μm)
496	501	24
488	496	14
488	501	10
496	514	7
488	514	5

Thirdly, the fringe pattern does not indicate the sign of the gradient of d – for example, a series of closed loop fringes may correspond to a 'hill' or a 'hollow'. Two possible methods of determining this quantity are outlined below. The first is a general technique, the second can be applied only in certain cases. Similar methods are used in the interpretation of fringes in 'classical' interferometry (18).

In the first method the direction of motion of the fringes is observed when a small displacement is applied to the object whilst the fringes are being viewed. (This is analogous to the method described in 2.4.4 for finding the sign of the gradient of the displacement in holographic interferometry).

It can be shown (17) that when the shape-difference fringes are observed using subtraction, the direction in which the fringes move depends on the sign of the gradient of the shape difference, on the direction of the displacement and on whether the wavelength of the light used in the stored speckle pattern is less or greater than the wavelength used in the 'live' pattern.

When the fringes are obtained by addition, the direction of fringe motion depends on the sign of the gradient of the shape difference and on the direction of the displacement.

The following 'fool's rule' can be applied in either case when:

    (i) the displacement is towards the viewing lens:
    (ii) the positive d direction points towards the viewing lens; and
    (iii) in the case of subtraction, the recording is made at the shorter wavelength.

The rule states that 'fringes roll down hills'. Thus a set of closed curves represent a hill if the fringes move outwards, and a hollow if they move inwards – see Figures 5.7 (a) and (b).

Figure 5.8 shows contour fringes obtained by comparing a flat specimen, which has a ramp milled out, with a flat master wavefront. The direction of the fringe motion is indicated by the arrow, showing that the ramp is descending with respect to the plane in this direction.

The second technique for determining the sign of the gradient of the shape difference requires the introduction of a tilt into the reference state: the master component will then give a fringe pattern representing this tilt and a test component will give a fringe pattern representing the difference between the shape-difference fringes and the tilt fringes.

If the shape difference consists predominantly of a rotation of the surface about a particular axis with respect to the master surface, and if the tilt of the master is about an axis orthogonal to that axis, the sign

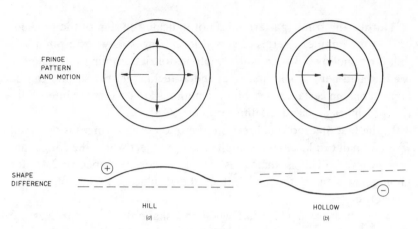

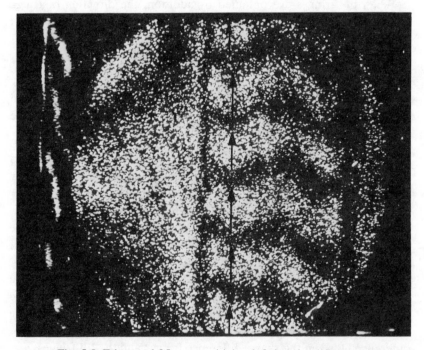

FRINGE
PATTERN
AND MOTION

SHAPE
DIFFERENCE

HILL
(a)

HOLLOW
(b)

Fig. 5.7 (a) Fringe motion observed for the viewing convention discussed in Section 5.4.2 when the shape difference is a 'hill'. (b) Fringe motion observed for the viewing convention discussed in Section 5.4.2 when the shape difference is a 'hollow'

Fig. 5.8 Fringes of 25 μm sensitivity defining the difference between a plane wavefront and a flat surface in which a ramp has been milled. (The arrows indicate the direction of fringe motion observed in accordance with the convention discussed in Section 5.4.2)

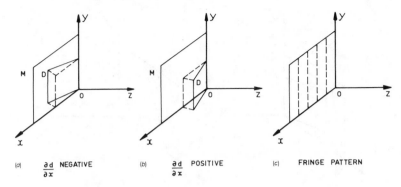

Fig. 5.9 (a) Test component, D, tilted about y-axis with respect to the master, M, $\partial d/\partial x$ negative. (b) Test component tilted about y-axis with respect to the master, M, $\partial d/\partial x$ positive. (c) Fringe pattern obtained for states (a) and (b) when master is a zero fringe reference state

of the gradient of the shape difference can be inferred from the fringe pattern obtained; this is seen as follows.

In Figures 5.9(a) and (b), the master surface lies in the xy-plane and the test surface is shown having a rotation about the y-axis with ($\partial d/\partial x$) being negative and positive respectively. With a zero fringe reference state, the fringe pattern shown in Figure 5.9(c) will be obtained in each case.

In Figures 5.10 and 5.11, the master state is shown rotated about the x-axis with positive ($\partial d/\partial y$) for the object shown in Figure 5.9(a) and (b). The fringe patterns obtained represent the loci of constant difference as measured in the z-direction between the test and master surfaces. It can be seen (readily by those who can think in three dimensions, with more difficulty, perhaps with the assistance of visual aids such as two cards by those who cannot!) that the loci of constant difference slope to the right – i.e. in the negative x-direction for ($\partial d/\partial x$) negative, and to the left, i.e. in the positive x-direction for ($\partial d/\partial x$) positive. Hence the sign of the gradient can be determined from the orientation of the fringes.

It can be seen that if the number of fringes, N_m, introduced into the reference state is very much less than the number of shape-difference fringes, n_x, the departure from the vertical direction will not be observable. Similarly, if N_m is much greater than n_x, the departure of the fringes from the horizontal direction will not be measurable. It can then be seen that if the position of a fringe can be measured to a fraction q of a fringe

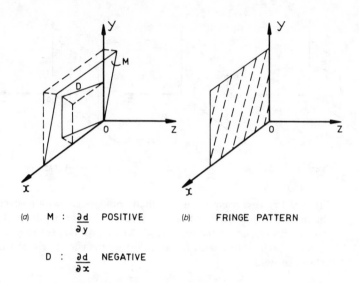

(a)   M : $\dfrac{\partial d}{\partial y}$ POSITIVE   (b)   FRINGE PATTERN

D : $\dfrac{\partial d}{\partial x}$ NEGATIVE

Fig. 5.10 (a) Test component D, oriented as shown in Figure 5.9(a) together with master, M, tilted about the x-axis. (b) Fringe pattern corresponding to the master and object orientations shown in (a)

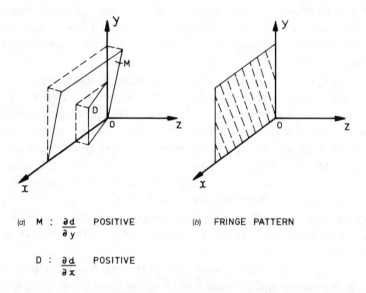

(a)   M : $\dfrac{\partial d}{\partial y}$ POSITIVE   (b)   FRINGE PATTERN

D : $\dfrac{\partial d}{\partial x}$ POSITIVE

Fig. 5.11 (a) Test component D oriented as shown in Figure 5.9(b) with master M tilted about the x-axis. (b) Fringe pattern corresponding to the master and object orientation shown in (a). (Note difference in fringe shape between this pattern and that shown in Figure 5.10(b))

spacing, the value of n_x must fall within limits

$$(n_x)_{\max} < \frac{1}{q}N_{\mathrm{m}} \qquad\qquad (5.26a)$$

$$(n_x)_{\min} < qN_{\mathrm{m}} \qquad\qquad (5.26b)$$

if the sign of the gradient is to be determined in this way.

It can be seen that if the shape difference has a component of rotation about the y-axis, the fringe pattern obtained with a rotated master reference state cannot in general be interpreted unambiguously to give the sign of the gradient of the shape difference in the y-direction.

In Figure 5.12 the two test surface positions D_1 and D_2 will give identical fringe patterns when the master state is represented by M. If, however, it can be assumed that $0 < n_y < N_{\mathrm{m}}$ where n_y is the number of fringes due to the shape difference in the y-direction, the sign of the gradient n_y and n_x can be determined when a known tilt is applied in the y-direction, and equations (5.26) are satisfied. This method of finding the gradient of the shape difference has the advantage that only one view of the fringe pattern is required to give both the shape difference and its gradient, while the first technique requires the comparison of two fringe patterns.

We have seen in this section that the absolute difference in shape between test and master surfaces can be found by finding:

   (i) the value of θ over the whole surface;
  (ii) a single value of the absolute value of d;
 (iii) the fringe positions;
 (iv) the sign of the gradient of the shape difference.

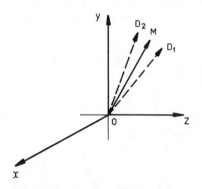

Fig. 5.12 An example of master and inspection component orientations for which the fringe orientation technique of gradient determination will break down

5.4.3 *Errors in ESPI shape measurement*

The correlation term S_{corr} in equation (5.22) is that which is obtained in an ideal system where a component which matches the master shape gives a null fringe pattern, and a non-matching component gives fringes which map out the difference between the test and the master component. In this section, we will consider how errors may arise and how they will affect the fringe patterns obtained.

True difference fringes are obtained only when the term $(\psi_1 - \psi_2)$ is constant across the image plane. If ψ_1 and/or ψ_2 are not constant across the image plane, the fringe patterns will represent both shape difference and errors due to these phase variations. If we write

$$\psi_1 = \psi_1{}^0 + \varepsilon_1 \tag{5.27a}$$

and

$$\psi_2 = \psi_2{}^0 + \varepsilon_2 \tag{5.27b}$$

where $\psi_1{}^0$ and $\psi_2{}^0$ are constant and ε_1 and ε_2 vary with position in the image, we have

$$S_{corr} = K' \cos^2 (\sin^2)$$
$$\times \left[\tfrac{1}{2}(\psi_1{}^0 - \psi_2{}^0) + 2\pi \left(\frac{1}{\lambda_1} - \frac{1}{\lambda_2} \right) d \cos \theta + \frac{\pi}{\lambda_1}\varepsilon_1 + \frac{\pi}{\lambda_2}\varepsilon_2 \right] \tag{5.28}$$

where

$$K' = K \langle I_R \rangle \langle I_s \rangle = \text{constant}$$

If ε_1 and/or ε_2 vary across the image, the fringe pattern obtained will not represent true difference fringes but will also represent the variation of ε_1 and ε_2.

If we write

$$\varepsilon_1 = \varepsilon \tag{5.29a}$$

$$\varepsilon_2 = \varepsilon + \Delta\varepsilon \tag{5.29b}$$

we than have

$$S_{corr} = K' \cos^2 (\sin^2)$$
$$\times \left[\tfrac{1}{2}(\psi_1{}^0 - \psi_2{}^0) + 2\pi \left(\frac{1}{\lambda_1} - \frac{1}{\lambda_2} \right) (d \cos \theta + \varepsilon) + \frac{2\pi}{\lambda_2}\Delta\varepsilon \right] \tag{5.30}$$

Variations in $\Delta\varepsilon$ occurring at one wavelength only give wavelength-sensitive error fringes, while variations in ε common to both wavelengths gives reduced sensitivity fringes of Λ sensitivity. Ideally, it should be ensured that such errors have magnitudes of only a fraction of a fringe: it is seen that error phase variations common to both wavelengths are less serious than those occurring at only one wavelength.

If such errors cannot be avoided, the system may nonetheless be used to measure shape difference provided that the errors are known. The errors can be found by inspecting a component of known shape in the system and comparing the fringe pattern obtained with the pattern which would be observed in the absence of errors.

Sources of error and their elimination are discussed in Sections 5.5.4 and 5.6.1.

5.4.4 *Linking adjacent views of a component*

It is not always possible to inspect the whole of one component in one view. This may be because its area is too large for the available master wavefronts, or because the shape difference involved would give rise to too many fringes to be resolved by the television system (Section 4.3). When this is the case, it is possible to inspect the test component in sections and link the measurements from these sections to give the shape difference information for the whole surface. For this sort of measurement a component of the same size as the test component which either matches the master shape, or whose departure from this shape is known, must be available. The procedure is as follows.

The component which will be called the master component, should be marked out by a coordinate grid with minimum spacing considerably less than the size of the sections being inspected. A kinematically relocating jig with suitable reference points in which the master component and then the test component may be located is also required.

Initially, the master component is located in the jig which is placed so that the first section is illuminated by the master wavefront. If the master component has the same shape as the master wavefront, the jig position and orientation are adjusted to give zero, or in some cases straight line, fringes (see Section 5.4.2). If the master component does not have the same shape as the master wavefront, the jig must be adjusted until the fringe pattern represents the difference between the master wavefront and the master component. The grid positions are noted.

The master component is now removed from the jig, which must not be moved, and the test component is inserted. The fringe pattern obtained gives the shape difference between the test component and the master wavefront.

The master component is reinserted and the jig is moved so that an adjacent section of the component is illuminated. It is desirable, though not necessary, that there should be some overlap between adjacent sections. The jig is adjusted to give either a null pattern or the fringe pattern specified by the known shape of the master. The grid positions are again noted. The test component is inserted and the shape-difference fringes recorded.

We now consider how the two fringe patterns obtained are linked together.

Each fringe pattern enables a function $\xi(r)$ to be measured which represents the variation in the difference in shape between the test and the master wavefront. The function $\xi(r)$ is related to the absolute shape difference by an unknown constant. Because the jig is moved between the two views, this constant is not necessarily the same for the two views. If we represent the relative shape difference measured by the first and second views as $\xi_1(r)$ and $\xi_2(r)$ and the absolute shape difference by $\xi_0(r)$, we have

$$\xi_0(r) = \xi_1(r) + C_1 \tag{5.31a}$$

$$\xi_0(r) = \xi_2(r) + C_2 \tag{5.31b}$$

where C_1 and C_2 are unknown constants.

If the two views do not overlap then the functions $\xi_1(r)$ and $\xi_2(r)$ must be extrapolated to give values of $\xi(r)$ in the overlapping region. A value D_{12} is assigned to $(C_1 - C_2)$, which minimizes the value of $[\xi_1(r) - \xi(r)]$ in this region. The value of C_1 must be measured independently and the absolute shape difference in the first section is now given by

$$\xi_0(r) = \xi_1(r) + C_1 \tag{5.32a}$$

and in the second section by

$$\xi_0(r) = \xi_2(r) + C_1 + D_{12} \tag{5.32b}$$

The shape difference in the nth section is then given by

$$\xi_0(r) = \xi_n(r) + C_1 + D_{12} + D_{23} + \cdots + D_{n-1,n} \tag{5.33}$$

This technique can in principle be used to measure the absolute shape of a component by comparing it section by section with a flat master wavefront. In practice it is useful only if the surface does not vary too drastically from planar.

5.4.5 *Simultaneous two-wavelength illumination*

As noted in Section 5.4.1, shape-difference fringes can be obtained by addition, avoiding the use of a video store, if the object is illuminated simultaneously by master wavefronts at two wavelengths. To do this, the two beams must propagate in the same direction simultaneously. A fairly simple, though not entirely reliable way of doing this using an Argon laser with an etalon is outlined below.

When the front face of the etalon is aligned normal to the laser beam in the cavity, a second cavity is set up in the tube which produces another laser beam. This beam generally runs only at 488 nm. Thus if the laser is tuned to one of the other Argon lines, the output will contain both these wavelengths travelling in the same direction. The procedure for aligning the etalon normal to the laser beam is usually given in the laser handbook as it is part of the normal laser alignment procedure.

To get good contrast shape difference fringes, the power in the two lines must be roughly the same. This is not very easily achieved using this technique and requires careful balancing in input electrical current and the output aperture of the laser. A neutral density filter is generally required in the output beam as it may be necessary to run at a higher power than is required optically to get the correct balance.

5.5 ESPI shape measurement using wavefronts generated by conventional optical components

It was shown in 5.4.1 that a laser beam which has been expanded and collimated forms a flat master wavefront. Spherical master wavefronts, both concave and convex, can be produced using lenses, and a cylindrical master wavefront can be made using lenses and a cone. These arrangements are shown in Figures 5.13(*a*), (*b*) and (*c*) respectively. Components whose surfaces are nominally flat, spherical or cylindrical can therefore be compared with these wavefronts using ESPI.

5.5.1 *Theoretical principles*

To obtain shape-difference fringes, the point to which the master wavefront converges must be located at the centre of the viewing lens of an ESPI system. This can be done using either the beamsplitter or the hole-in-the-mirror type reference beam systems discussed in Section 4.5.1. This is considered in more detail in the next section.

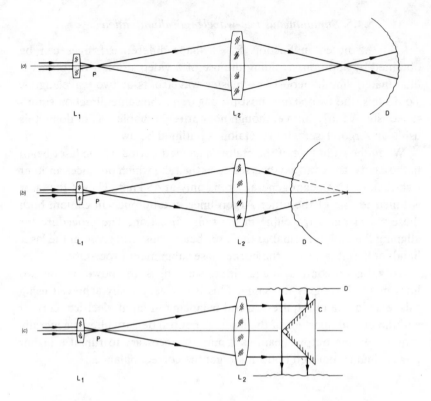

Fig. 5.13 (*a*) The formation of master spherical wavefront
for the inspection of a concave spherical surface. (*b*) The formation
of a master spherical wavefront for the inspection of a convex
spherical surface. (*c*) The formation of cylindrical master wavefront.
C is a right angled front reflecting cone

If the departure of the surface shape from the master wavefront shape
at a given point is given by d, the speckle correlation term is given by
equation (5.22) as

$$S_{\text{corr}} = K' \cos^2(\sin^2) \left[\tfrac{1}{2}(\psi_1 - \psi_2) + 2\pi \left(\frac{1}{\lambda_1} - \frac{1}{\lambda_2} \right) d \right] \qquad (5.34)$$

In this case, $\cos \theta = 1$ for the whole object, since the illumination is
normal at all points. The shape difference between test and master
shapes is therefore mapped out at intervals of:

$$\Lambda = \frac{\lambda_1 \lambda_2}{2(\lambda_1 - \lambda_2)} \qquad (5.35)$$

It should be noted that when flats, spheres or cylinders are being inspected by this technique an absolute shape is not defined (the plane, spherical and cylindrical waves maintain their shape over considerable distances) so that a single position measurement in the case of a flat, or a radius measurement in the case of spheres and cylinders, is necessary to specify the absolute position of the test surface.

5.5.2 *Optical arrangements*

Two arrangements which enable the point of convergence of the master wavefront to be located at the centre of the viewing lens aperture of an ESPI system are shown in Figures 5.14(*a*) and (*b*).

In Figure 5.14(*a*), the laser beam is expanded by the lens L_1 and partially passes through the beamsplitter, B, into the master wavefront optics. The returned beam is partly reflected by the beam splitter onto the lens, L, which is the viewing lens of an ESPI. This may be a system where the smooth reference beam is added by using a beamsplitter (Figure 4.8), or by a hole-in-the-mirror arrangement (Figure 4.9).

In Figure 5.14(*b*) the master wavefront is obtained by expanding the beam with the lens L_1 and then sending it through a small hole in the mirror M into the master wavefront optics. The returned light is reflected by the mirror, M, onto the lens, L, which again is the viewing lens of an ESPI. This arrangement is identical to that of Figure 4.10 if the speckle reference beam is introduced using the hole-in-the-mirror arrangement.

When a beamsplitter is used in the object beam, a lot of the input light is wasted (at least 75% if a 50/50 beamsplitter is used), so that, if possible, the hole-in-the-mirror arrangement should be employed. When very high sensitivity contours are being observed, however, the departure from conjugacy with this system may give rise to error fringes.

The loss of light which occurs when a beamsplitter is used to introduce the speckle reference beam is not a significant consideration since the fraction of the input laser power which goes into the reference beam is very small. However, dust particles on the beamsplitter will give rise to additional noise in the image which will reduce the contrast of the fringes even when subtraction is used to obtain the fringes, since the noise changes when the wavelength changes. Thus, a hole-in-the-mirror arrangement is preferable except when high sensitivity contours are being observed, when the departure from conjugacy may again give rise to error fringes.

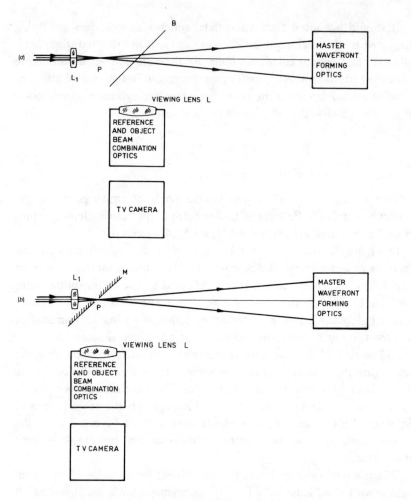

Fig. 5.14 (*a*) Conjugate object-beam illumination geometry obtained using a beamsplitter element, B. (*b*) Near-conjugate object-beam illumination geometry obtained using a hole-in-mirror element, M

5.5.3 *Alignment procedure*

It is necessary to use a component whose shape is known to align the shape measuring system: the procedure for alignment is as follows.

The ESPI is aligned to give an in-line object beam. The lens which expands the object beam in the interferometer corresponds to the lens L_1 in the arrangements of Figures 5.13(*a*), (*b*) and (*c*). The component(s) forming the master wavefront must then be placed in approximately their correct positions, and the master component is placed in its jig so that it is illuminated by the object wavefront. A dim image of the component should be seen on the television screen. The position and orientation of the master component must then be adjusted to have maximum brightness – this will occur when the light is being reflected in approximately the specular direction, indicating that the master wavefront and component are roughly aligned. Under these conditions, the system should be sufficiently in alignment to give correlation fringes. The object and the wavefront forming optics should now be manipulated to give a null fringe if the object matches the test component, or the appropriate shape-difference pattern if they are different. This procedure is very much simpler when the addition process is used to see the fringes; when subtraction is used, the procedure requires care and patience.

When a test component is substituted in the jig, correlation fringes are obtained which represent the difference between it and the master wavefront.

5.5.4 *Error fringes in shape-difference measurement for conventionally generated master wavefronts*

Error fringes are obtained in this system if any of the following conditions occur:

(i) The optics producing the master wavefront are inadequate (e.g. when a lens having considerable spherical aberration is used to produce a 'flat' master wavefront).

(ii) The optical components are adequate but are misaligned.

(iii) The jig holding the test component is not aligned properly with respect to the master wavefront (see Section 6.9.2).

The first condition should not arise if diffraction-limited optics are used. If a component of known shape is available, it may be used to calibrate the system.

The second and third conditions will be avoided if the procedure outlined in the previous section is followed.

If, however, a known component is not available, relative shape differences between one component and another may still be determined.

5.6 The inspection of components of complex shape using holographically generated master wavefronts

Master wavefronts of complex shape can be produced if a component of the master shape is available which has a specularly reflecting (mirror) finish. The way in which the wavefront is produced, the practical difficulties in realizing the ideal system, and the means of overcoming these difficulties are discussed in Section 5.6.1. Section 5.6.2 shows how the wavefront can be used to compare the shape of nominally identical components with the master shape.

5.6.1 *The manufacture of the master holographic element*

The specularly reflecting master component M – see Figure 5.15(a) – is illuminated by a diverging spherical wavefront. The light reflected from the surface is collected on the hologram plate, H, which is also illuminated by the holographic reference beam, U_r. The plate is developed and relocated in its mount. When the plate is reilluminated by a beam which is conjugate to the original reference beam (i.e. travelling in the opposite direction at all points in the beam), a beam is reconstructed which is conjugate to the beam which was reflected by the master component M. This beam will be reflected by the latter and will converge to the point from which the original object beam diverged – see Figure 5.15(b). When such wavefronts are made at two wavelengths, they may then be used as master wavefronts to compare the shape of test components with that of the master component. If the reconstructed beam is not accurately conjugate with the original object beam, it does not constitute an accurate master wavefront, and it is likely that error phase terms will appear when a component is inspected.

In order that the reconstructed object beam is accurately conjugate with the original object beam, several conditions must be satisfied.

(i) The reconstructing reference beam must be accurately conjugate with the original holographic reference beam.

(ii) The holographic recording medium must not change shape or size, and it must be returned to its original position with respect to the master component.

(iii) The two faces of the plate on which the holographic recording medium is laid (generally a piece of glass) must be optically flat. They need not, however, be parallel to one another.

The first condition is most readily satisfied by using a single aberration-free plane wavefront for both making and reconstructing the hologram. Such a system is considerably easier to align than one where converging and diverging spherical wavefronts are used.

The second condition is satisfied if a suitable relocating mount is used for the hologram plate (see Section 6.4.2) and if the emulsion is preshrunk (see Section 6.5.2).

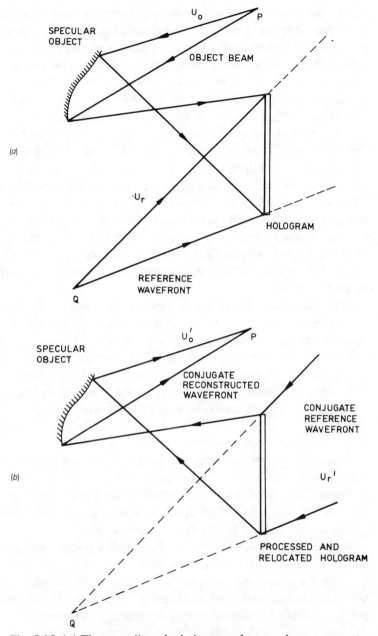

Fig. 5.15 (a) The recording of a hologram of a specular component, M. (b) The reconstruction of the conjugate real wavefront from the hologram recorded as shown in Figure 5.15(a). (Note that this is focussed to a point P which is coincident with the original point of divergence of the object illumination wavefront)

The third condition is not, however, readily satisfied since commercially available plates coated with holographic recording materials are not generally optically flat, and coating plates with holographic materials is not easy. In certain circumstances, this condition can be relaxed without affecting the ability of the master wavefront to perform the shape measurement. This will be seen by considering in more detail the effect of non-flat hologram plates on the master wavefront.

When the hologram is reconstructed by the conjugate reference beam, the holographic reference beam passes through the hologram plate before illuminating the hologram. If the two faces of the hologram plate are flat and mutually parallel, the beam striking the emulsion will be unchanged in shape (see 1.5.2) and the conjugate object beam is reconstructed accurately. If the two faces of the hologram plate are optically flat but inclined at an angle to one another, the orientation of the emerging beam is altered but its shape is unchanged, so that a suitably oriented input beam will give again an accurate conjugate reconstruction of the object beam.

When, however, the relative orientation of the two faces varies across the hologram plate, the transmitted beam will no longer have the same shape as the input beam, since the angle at which a given ray emerges depends on the relative orientation of the two faces at the points at which that ray enters and leaves the plate. Figures 5.16 (*a*) and (*b*) show two extreme examples. In Figure 5.16(*a*) the two faces are parallel, and the emerging ray is at the same angle as the incident ray. In the second case the two faces are converging sharply, and the angle between the directions of the incident and emergent rays is large.

When such variations of face orientation are present, the reconstructing reference beam is no longer accurately conjugate to the original reference beam, and the reconstructed object beam will be distorted. The amount of the distortion depends on the curvature of the original object wavefront. This can be seen as follows.

Because the object wavefront in this case is a smooth wavefront, the hologram will consist of a grating whose spacing varies as the angle between the object and reference beam varies. If the original reference and object beams were incident at angles θ and ψ respectively at a given point, the grating spacing d at that point is given by equation (1.72) as

$$d = \lambda/(\sin \theta + \sin \psi)$$

When the hologram is reconstructed at an angle $(\theta + \Delta\theta)$, the reconstructed object beam will be diffracted into an angle $(\psi + \Delta\psi)$ where

$$\Delta\psi = \frac{\sin \theta}{\sin \psi}\Delta\theta \tag{5.36}$$

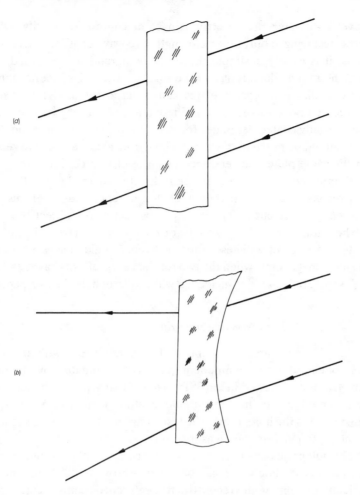

Fig. 5.16 (*a*) The ray path of the holography reconstruction
wavefront passing through a plate having optically flat parallel faces.
(*b*) The ray path of the hologram reconstruction wavefront passing
through a plate of varying thickness. This introduces distortions into
the reconstructed wavefront

Thus if θ does not vary very much over the hologram (i.e. the object
beam is not very sharply curved), then the variation in $\Delta\psi$ is considerably
less than if the values of θ cover a wide range, so that the distortion in
the former case is much less than in the latter. Hence the sharper the
curvature of the component, the greater the distortion.

This distortion in the master wavefront will give rise to error phase
terms of the form given in equation (5.28). It was seen in (5.30) that
error terms which are common to the master wavefronts at the two

wavelengths give rise to error fringes of contour sensitivity, whereas those occurring at only one wavelength give wavelength sensitive fringes. Thus, it is clearly desirable that both holograms are recorded on the same hologram plate. It has been found that when this is done, commercially available hologram plates can be used for moderately sharply curved components such as turbine blades, but it may be necessary to use specially coated flat plates for very sharply curved components.

It should be noted that when the holograms at λ_1 and λ_2 are recorded on the same plate, the reference beam used to record and reconstruct at λ_1 must be incident in a different direction from that used to record and reconstruct at λ_2. If only a single reference beam were used, the reference wavefront at λ_1 would reconstruct a true object beam from the hologram made at λ_1 and a distorted object beam from the hologram at λ_2 and vice versa. These spurious beams would give rise to noise in the final image. Only when the two reference wavefronts have reasonably well separated angles of incidence will this 'crosstalk' be avoided.

5.6.2 *The inspection procedure*

To compare the shape of test components with the master component, the test component must be illuminated by the master wavefronts and viewed in an ESPI. Thus, the master component should be mounted in a jig having suitable locating points (see Section 6.4.2) when making the holograms. The test component can then be substituted in the jig to perform the measurement.

The hologram plate mount and master component jig should be rigidly fixed to a base plate which can be rotated through 180° to minimize the difficulty of aligning the reconstructing reference beams accurately. The reference beams should be directed onto the hologram plate by means of mirrors having tilt and rotation adjustment for fine adjustment of the beam directions. When the hologram plate has been developed and relocated, the base plate is rotated through approximately 180°. The reconstructed object beam is reflected by the master component, and the orientation of the base plate is adjusted to minimize the spot size of this reflected beam at the point corresponding to the original point of divergence of the object beam.

The viewing lens of the speckle interferometer which may have a hole in the mirror or a beamsplitter reference beam is placed at this point so that the light is normally incident on the lens (Figure 5.14). An image of the master component will now be seen on the television monitor, and since its surface is specular, it combines with the speckle reference beam to give an interference pattern. The fringe patterns at the two wavelengths give rise to Moiré fringes when they are compared by addition or by subtraction. At this stage the directions of the reference beams are adjusted to give a null Moiré fringe pattern. When this has been achieved a test component may be substituted for the master component, and shape difference fringes will be obtained of sensitivity Λ given by

$$\Lambda = \frac{\lambda_1 \lambda_2}{2(\lambda_1 - \lambda_2)} \frac{1}{\cos \theta} \tag{5.37}$$

If null fringes cannot be obtained on the master component, as will occur, for example, if the hologram plate is not sufficiently flat, the shape difference between test and master component can nonetheless be obtained. The fringe pattern obtained on the master component is noted, and the 'error' shape difference implied by this pattern is subtracted from the shape difference given by the test component fringe pattern, to give true shape differences.

Finally, it should be noted that any departure of the finish of the master component from specular will reduce, and very quickly lose, the fringe visibility of the shape-difference fringes.

Results which show how the technique has been used in the inspection of a turbine blade profile are described in Section 7.6.

5.7 Limiting factors in ESPI shape measurement

5.7.1 *Range of sensitivity*

The difference in shape defined by a fringe in the ESPI technique is determined by the difference between the two wavelengths used. Using an Argon laser, the fringe sensitivity can have various values between 2 μm and 30 μm (see Table 5.1). With a Krypton laser, the sensitivity may have values between 1 and 50 μm, and if a suitable dye laser is used the range can be varied from 1 μm up to tens of millimetres.

The overall shape difference must be such that the spacing of the fringes does not exceed the limit specified in Section 4.3 – i.e. approximately 1/100th of the field of view. Shape-difference fringes are generally quite complex in form and very fine fringes will be more difficult to follow than straight fringes of the same spacing, so that the limit can be even lower than this. This effect is compounded by the fact that the fringes may have reduced visibility due to speckle pattern decorrelation – see Section 5.7.3 and Appendix F.

5.7.2 *Overall area*

In the case of surfaces which are nominally flat, spherical or cylindrical the maximum area which can be inspected in one view is governed by the optical components available. Inspection of a flat surface of a given size requires a precision collimating lens of the same size. The length of cylinder which can be inspected in one view depends on the size of the collimating lens and the cone used.

The fraction of a concave spherical surface which can be inspected in one view is determined by the numerical aperture of the lens used to produce the spherical wavefront. The intensity of the image obtained depends on the radius of the hemisphere and on the diameter of the wavefront forming lens.

A standard quality double achromat lens will give an accurate spherical wavefront subtending an angle of ≃25° together with a reasonable image intensity. If it is required to extend the angle beyond this value, either a microscope objective which has a small aperture and hence gives a dim image, or a specially manufactured aspheric lens must be used.

In the case of convex spherical surfaces, the area which can be inspected in one view is somewhat less than the area of the lens used to produce the required converging spherical wavefront, and the angle subtended by the spherical section viewed is determined by the numerical aperture of the lens.

Whether the whole of a surface of a particular component can be inspected using ESPI in conjunction with a holographic master element (Section 5.6) depends on whether or not the light reflected from a mirror finish master component can be collected onto an area for which suitable holographic reference beams are available. It was seen in Section 5.6.1 that these reference beams should be flat to better than a wavelength. This means that if the light reflected by the master is spread over a sizeable area, large diameter, high quality collimating lenses must be used. It is possible to use diverging and converging beams for the making and reconstructing of the holograms but in this case the size of the lenses used to produce the converging beams has to be greater than the area of the object beam, and the alignment becomes quite difficult.

The more sharply curved the component, the more it spreads the light out and the area of the master hologram increases accordingly. The size of hologram required to inspect a given component can be found most easily by observing the light reflected from a mirror finish component (not necessarily a master). If such a component is not available, the form of the reflected light must be investigated by ray tracing.

5.7.3 Surface finish restrictions

It is seen in Appendix F that the speckle pattern obtained from a given surface changes as the illuminating wavelength changes, and that the rate at which it changes is a function of the surface roughness as well as of the change in wavelength. An expression is given for the upper limit to the standard deviation of the surface height, σ_s, which will give reasonable visibility fringes at a given contour sensitivity, Λ, for a master wavefront incident at an angle θ. This is

$$\left(\frac{1}{\lambda_1} - \frac{1}{\lambda_2}\right)^{-1} < 16\sigma_s \cos \sigma$$

which gives

$$8\sigma_s < \Lambda \tag{5.38}$$

This restriction applies to any form of ESPI shape measurement.

There is also a lower limit to the surface roughness which allows shape measurement by ESPI. Clearly, when the surface is sufficiently smooth, the specularly reflected component rather than the speckle component of the scattered light predominates, and a speckled image is not obtained. The addition of reference beam and object beam will then give rise to an interference pattern, rather than a speckle pattern. This in itself does not prevent the shape measurement being made since Moiré fringes may be obtained by comparing the interference patterns at the two wavelengths (for example, Section 6.7.2). However, the image intensity will not be uniform if the semi-angle of the scattering cross-section as defined in Section 6.3.3 is small. This can be seen as follows.

In Figure 5.17(a) the test and master surfaces are parallel. In this case all the light in the scattering cone is collected by the viewing lens. In Figure 5.17(b) the test surface is tilted with respect to the master surface. In this case, the light in the scattering cone is not collected by the viewing lens. Thus, as the slope of the surface varies across the object, the amount of light in the image varies and some parts of the object may not be seen at all. This non-uniformity will degrade the visibility of the fringes and in some cases they will vanish altogether.

This effect cannot easily be quantified since it depends on:

> (i) the angle subtended by the viewing lens at the object surface;
> (ii) the size of the scattering cone of the surface; and
> (iii) the difference in slope between test and master surfaces.

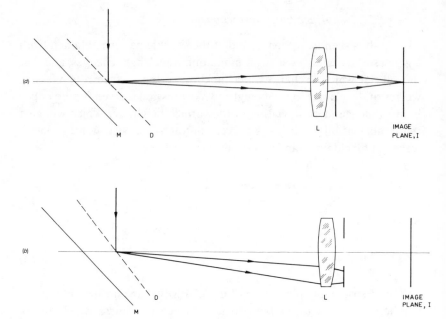

Fig. 5.17 (*a*) Ray path of a cone of light reflected from a specular (or quasi-specular) inspection object, D, from a region where its plane is parallel to that of the master, M. (*b*) Ray path of a cone of light reflected from a specular (or quasi-specular) inspection object, D, from a region where its plane is not parallel to that of the master, M

However, any surface which is predominantly a specular reflector is likely to cause problems unless the difference in shape is very small.

Thus, in some cases, the surface may be too smooth to measure its shape using ESPI, and in other cases it may be too rough. Silver spray paint has been found to give a finish suitable for the observation of shape difference fringes up to a sensitivity of 10 μm.

5.8 The direct comparison of components using two-wavelength ESPI

It is possible to compare the shape of two similar components directly using the ESPI system (19). An arrangement for doing this is shown in Figure 5.18. A wavefront is divided by the beamsplitter, B, so that it illuminates the two components, D_1 and D_2, which are situated so that their images are superimposed on the television camera at I. This can be done approximately by looking into the beamsplitter and

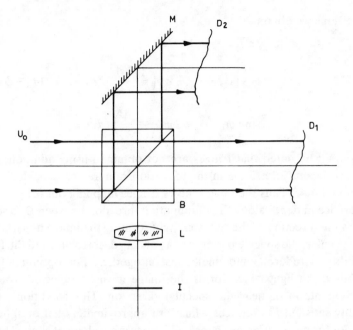

Fig. 5.18 A beamsplitting cube arrangement for the direct
comparison of two nominally identical components, D_1 and D_2

adjusting the position of one of the components to obtain zero parallax
between the two images.

The two components are illuminated at λ_1 and the intensity at the
camera is given by

$$\mathscr{I}_1 = I_1 + I_2 + 2\sqrt{I_1 I_2}\cos\left[\phi_{s_1} - \phi_{s_2} + \frac{4\pi}{\lambda_1}(d_1 - d_2) + \phi_1\right] \quad (5.39)$$

where ϕ_1 is the phase of the illuminating wavefront, and d_1, d_2 are the
total optical paths from the beamsplitter to the image plane via D_1 and
D_2 respectively, and I_1, ϕ_{s_1}, I_2, ϕ_{s_2} are the intensities and phases of the
speckle in the images of the two components.

When the illuminating wavelength is changed to λ_2, the intensity
becomes

$$\mathscr{I}_2 = I_1 + I_2 + 2\sqrt{I_1 I_2}\cos\left[\phi_{s_1} - \phi_{s_2} + \frac{4\pi}{\lambda_2}(d_1 - d_2) + \phi_2\right] \quad (5.40)$$

The above equation will apply when the light which forms the image of
the component is reflected from the components in or near the specular
direction, and the surfaces are not too rough (see the previous section
and Appendix F). When the two patterns are subtracted under these

conditions we obtain

$$\mathcal{I}_1 - \mathcal{I}_2 = 2\sqrt{I_1 I_2}$$

$$\times \cos\left[\phi_{s_1} + \phi_{s_2} + 2\pi\left(\frac{1}{\lambda_1} + \frac{1}{\lambda_2}\right)(d_1 - d_2) + (\phi_1 + \phi_2)\right]$$

$$\times \sin\left[\phi_1 - \phi_2 + 2\pi\left(\frac{1}{\lambda_1} - \frac{1}{\lambda_2}\right)(d_1 - d_2)\right] \qquad (5.41)$$

(It should be noted that fringes are not obtained using addition, since the two beams interfering in the television camera are 'speckled' – see Section 4.4.4.) Thus the image shows fringes mapping out $(d_1 - d_2)$, the difference in depth along the illumination direction between D_1 and D_2.

As the direction of the reflected light forming the image departs from the specular, the speckle patterns at λ_1 and λ_2 decorrelate, and the fringe visibility is reduced, and finally lost altogether. For optimum fringe visibility, the light which forms the image must be reflected from the components in or near the specular direction. This condition can be readily satisfied for surfaces which are approximately flat or spherical but not for more complex shapes. A holographic element similar to that discussed in Section 5.5 could possibly be used, but such an element would be located so far away from the test components that only a small fraction of the component could be illuminated by it.

Fringes that define the difference between two nominally plane surfaces have been obtained (19) and more recently the fairly flat portions of two aerofoil sections have been inspected using the same technique by one of the authors (C.W.).

The direct comparison technique is considerably less useful than the comparison technique discussed in Sections 5.5 and 5.6 due to: (a) the limitation in component curvature; (b) the inability to operate as an addition system; and (c) the doubling of decorrelation effects due to surface roughness. It could be used, however, for comparing surfaces which are nearly flat or spherical without having to produce a master component of specular finish.

5.9 Shape measurement using projected fringes

The non-localized interference pattern formed by the interference of two beams of light (for example, that obtained from a Michelson interferometer) may be used to form fringes on the surface of an object A fringe will appear each time the surface of the object intersects a fringe in the light beam. Figure 5.19 shows projected fringes on the

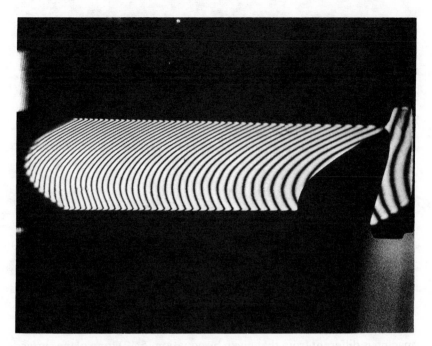

Fig. 5.19 Equispaced 'sheets' of fringes projected onto a section of curved aerofoil section

surface of a turbine blade aerofoil section. The shape of these fringes depends on the shape of the surface and also on the viewing and illumination directions. This is illustrated by the following simple example.

Assume that the fringes are formed by the interference of two plane waves, so that the fringes form mutually parallel planes of spacing L – see equation (1.26). A plane surface which is illuminated by the beams which are incident at a mean angle θ is viewed in the normal direction, see Figure 5.20. The projected fringes are straight lines of spacing q_f given by

$$q_f = \frac{L}{\cos \theta} \tag{5.42}$$

If part of the surface is tilted by a small angle γ with respect to the plane, the fringe spacing is now given by

$$q_f = \frac{L}{(\cos \theta - \gamma \sin \theta)} \tag{5.43}$$

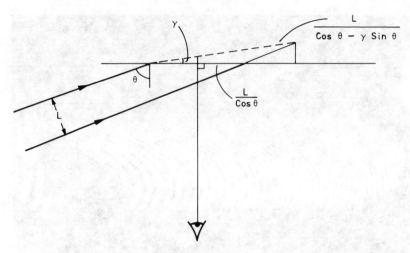

Fig. 5.20 The geometry of fringes of spacing L as projected onto a surface normal to the line of sight and inclined at a small angle, γ, to the line of sight

Thus, the departure of the surface from flatness gives fringes whose spacing is different from that given by equation (5.42) for a plane surface normal to the viewing direction, and the shape of the surface may be calculated from the fringe pattern.

If, however, the object is illuminated and viewed in the same direction (e.g. by using a beamsplitter to project the fringes), the fringes on the object surface will consist of straight lines of uniform separation regardless of the surface shape, so that the sensitivity to surface shape is zero in this configuration.

Maximum sensitivity is obtained when θ approaches 90°, and viewing is in the normal direction. Under these conditions, the surface must have a diffuse scattering finish in order to use the available light most efficiently. (Surfaces which are not plane will be partly in shadow when obliquely illuminated.)

While the fringes on the surface may be obtained by making a double exposure hologram where the orientation of the object illumination beam is altered between exposures, the resultant fringes are identical to those which would be observed if the object were simultaneously illuminated by two equivalent beams; hence the technique does not necessarily require holography. However, Abramson has shown (4) that the sandwich holography technique can be used to rotate the interference surfaces in any direction, which considerably simplifies the interpretation of the fringes. This technique is discussed in Section 5.9.1.

In Section 5.9.2, the use of projected fringes to compare nominally identical surfaces is discussed.

5.9.1 *Surface depth contouring using sandwich holography and projected fringes*

Sandwich holography, which is discussed in some detail in Section 6.6.2 is a technique developed by Abramson (4), by which a double exposure hologram may be tilted through quite small angles to give a large (up to 180°) rotation of the reconstructed fringes on the object.

The two exposures of the sandwich hologram are made with two object illumination beam directions, so that when it is reconstructed, projected fringes of the form discussed in the previous section will be seen on the object. When the hologram is tilted, the fringe pattern alters, and the fringes may be aligned to any required orientation. Figures 5.21(*a*) and (*b*) show results obtained by Abramson in such an experiment. The first photograph shows the fringes obtained when the original reference beam is used for reconstruction. In the second photograph the same hologram has been tilted so that the fringes are projected in planes perpendicular to the line of sight.

We see that a suitable tilt of the sandwich hologram allows the fringe surfaces to be aligned with respect to any required plane.

Abramson derives expressions (4) for the relationship between the rotation of the hologram and the rotation of the interference surfaces, but recommends that calibration techniques using reference objects of known dimensions and angles should be used for measurement purposes.

5.9.2 *Surface shape comparison using projected fringes* (5)

In this technique, fringes are projected onto an object surface, and a photographic image is made of the illuminated surface. If the developed photographic plate is relocated in its original position in the system, the live image of the object will cancel the negative image exactly. If the original component is then replaced by a component of different shape and illuminated by the original fringe field, the fringe pattern on the surface, and hence the fringe pattern on the image will be different. The two patterns in the image plane will 'beat' together to give so-called Moiré fringes, which will appear as a modulation of the amplitude of the projected fringes.

This modulation will appear as a coarser fringe pattern on the image and the spacing of the fringes is determined by the variation in surface

(a)

(b)

Fig. 5.21 (a) Holographic fringes projected onto an array of objects parallel to the direction of illumination. (b) Fringes projected on the same object as that in (a) but in a direction perpendicular to the line of sight. These fringes were obtained by applying a small tilt to the sandwich hologram and enable the depth of the object to be more readily determined. (Figures 5.21(a) and (b) are reproduced by kind permission of Dr Nils Abramson)

height, Δz, along the viewing direction. Fringes will occur when

$$\Delta z = \frac{L}{\sin \theta}$$

where L is the fringe spacing in the incident beam and θ is the angle of incidence of the projected fringe beam.

Thus the difference in shape between the components is mapped out at these intervals. The system is analogous to the ESPI system used in conjunction with a holographic element to record a master wavefront, which measures shape difference. It is however a less complex optical system requiring fewer optical components and only a single frequency laser.

It is important to note however, that the lens used to form the image of the master must be able to resolve the fringe pattern over the whole surface. As the depth of the object is increased, the depth of the focus of the lens must be increased by reducing its aperture. This will reduce the resolution of the lens so that the resolvable fringe spacing and hence the sensitivity is reduced.

In the case of flat surfaces, the upper limit to the resolution of the system is determined by the resolution of this lens, and can be of the order of 10 μm. This decreases as the curvature and hence the depth of the object surface is increased.

The requirement that the illuminating wavefront be incident at a very oblique angle means that the technique is useful only for components which are predominantly curved in one direction – it could not be used, for example, in the inspection of spherical surfaces.

6

Experimental design and technique

6.1 Introduction

The successful use of the techniques described in the previous chapters requires that some insight into experimental design and technique be gained. A good way of doing this is to actually carry out the required experiment together with the associated measurements. This approach is quite sound as long as it is borne in mind that a considerable amount of time and money can often be saved if the experiment is based on a brief theoretical study of the practical factors involved. The contents of this chapter are intended to provide the basis for such an approach. Readers may also find that the practical details discussed enhance their understanding of the theoretical principles already expounded.

6.2 Some factors affecting the selection of experimental technique

The techniques discussed in this book enable the following types of measurement to be made:

(i) Static and quasi-static surface displacements, using holographic interferometry (Chapter 2), or speckle pattern interferometry (Chapters 3 and 4).

(ii) Dynamic surface displacements using modified versions of the same general method as (i).

(iii) Surface shape based on dual wavelength Electronic Speckle Pattern Interferometry, dual wavelength holographic interferometry and fringe projection methods (Chapter 5).

Sensitivities for the above methods have been defined at various points in the text. The sensitivities of the various displacement techniques are summarized in Table 6.1 and the accompanying notes.

Table 6.1 *Displacement sensitivities*[a]

Technique	Component					
	d_1[b]	d_2, d_3[b]	$d_{1,2}, d_{1,3}$	$d_{2,2}, d_{2,3}$ $d_{3,3}, d_{3,2}$	$\dfrac{\partial(d_{1,j})}{\partial x_k}$, $j, k = 2, 3$	Independent observation[c]
Holographic interferometry (HI)[d]	10–10^3 μm	10–10^3 μm	0.3 μm per fringe	1 μm per fringe	No[e]	No
Double exposure SPP[d]	No	1–10^2 μm	10^{-3}–10^{-5} rad across the field of view[f]	1–10^2 μm per fringe	10^{-3}–10^{-5} rad per fringe	Yes
SPC + ESP interferometry[g]	No	No	0.3 μm per fringe[h]	0.4 μm per fringe	10^{-3}–10^{-5} rad per fringe	Yes

[a] Where only one number is quoted, the sensitivity may be varied by a small amount (less than an order of magnitude) by altering the illumination and viewing geometries. Where a range of sensitivies is quoted, this should not be taken as a rigid definition of the upper and lower limits, but as an indication of the range which can be covered in a typical viewing arrangement. It should be noted that when the displacement has components to which the system is not sensitive, these components may cause speckle pattern decorrelation which can degrade the visibility of the fringes.

[b] These columns represent rigid body translations, and the sensitivity here indicates the amount of movement required to give one fringe in the field of view.

[c] This column indicates whether or not the specified components may be measured independently.

[d] May be used to measure dynamic displacement for all displacement components detectable.

[e] This does not include indirect techniques such as Moiré fringe formation by the superposition of holographic fringe negatives.

[f] For the double exposure SPP techniques these results correspond to the fringe spacing obtained using the diffraction halo point by point measurement technique (Section 3.2.2).

[g] May be used to measure $d_{1,2}, d_{1,3}$ dynamic displacements only.

[h] Reduced sensitivity (up to 100 μm) fringes can be obtained using the arrangement discussed in Section 3.7.1.

This information can be used as a guide when a technique is to be selected for the solution of a particular problem. For example, in the case of a displacement measurement the requirements may be specified in terms of:

> (*a*) the direction and magnitude of the displacement to be measured;
> (*b*) the rate of displacement;
> (*c*) the area of surface to be observed in a single view;
> (*d*) the geometry of the surface and the nature of its finish (see for example Section 6.2).

(*a*) and (*b*) will usually enable the type of interferometer to be selected on the basis of the information already summarized in this section. (*b*), (*c*) and (*d*) will govern the basic interferometer design. The amount of laser power required is governed by (*c*) and (*d*) (see Section 6.3).

This approach enables one to establish an outline design for the system. Individual optical components may then be specified on the basis of the required beam expansions, viewing distances, image sizes etc. At this stage it is often useful to draw a scale layout and an approximate ray diagram of the interferometer.

6.3 Light intensity considerations in speckle pattern and holographic interferometry

When a recording is being made of an object in either speckle or holographic interferometry, a certain level of light energy is required in the recording plane. This level is determined by the properties of the recording medium. The amount of laser power necessary to produce the required recording-plane energy is determined by the size, shape and surface finish of the object and by the optical configuration. It is also affected by the mechanical stability of the system since this may determine the recording time.

In this section we will discuss the properties of photographic and holographic emulsions insofar as they affect the intensity requirements of holographic and speckle interferometers. In addition, the relationship between input laser power and recording-plane intensity is considered, so that the laser power required to perform a given measurement can be estimated.

In holographic and speckle correlation interferometry arrangements where a smooth reference beam is used, the object beam path attenuates the light by a very much greater amount than the reference beam path

so that the laser power required can be considered to be solely determined by the object beam path.

When two speckled beams are used in SPC interferometry then the
intensity requirements of both beams must be considered. In double
exposure SPP, only one beam is used. Thus, in determining laser power
requirements, we are primarily concerned with the amount of light which
is scattered by the object under inspection to either the holographic
recording plane or the image plane of a viewing lens.

6.3.1 *Light intensity characteristics of the photographic process associated with holographic and speckle pattern recording*

Conventional photographic emulsions consist of an intimate
distribution of silver halide crystals suspended in gelatine. When light
of sufficient intensity is incident on the emulsion, development centres
in the form of stable clusters of silver atoms are created. These centres
catalyse the deposition of silver when the emulsion is immersed in
developer. Unexposed regions are then made water soluble in the fixing
process and are removed by rinsing the emulsion in water. The net result
of this process is that the negative processed film attenuates the light
incident on it by an amount proportional to the light energy incident
originally. (At one extreme it can be seen that regions of the emulsion
where the exposure intensity is high will correspond to zones of maximum
silver deposition and hence maximum attentuation in the processed film.)

The exposure energy E is defined as

$$E = It_e \tag{6.1}$$

where I is the intensity of the light in the recording plane and t_e is the
exposure time.

The transmittance of the processed plane, T, is defined as

$$T = \frac{\text{Intensity of the light transmitted by processed plate}}{\text{Intensity of the light incident on processed plate}} \tag{6.2}$$

The transmittance is related to the exposure energy by a curve of the
general form of Figure 6.1.

It should be noted that equation (6.1) would suggest that a given
transmittance can be obtained for any level of incident light intensity
simply by adjusting the exposure time, but this is not the case: a specific
minimum intensity is required to generate stable development centres
for a given emulsion, and if the intensity falls below this value no

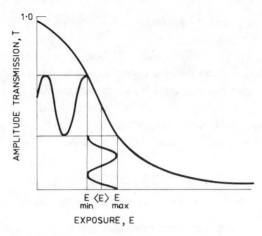

Fig. 6.1 The transmittance exposure characteristic of a photographic emulsion

recording is obtained however long the exposure time. This is known as reciprocity failure.

We will consider first how the T/E curve affects the recording intensity when making a hologram.

It was seen in Section 1.7.3 that an accurate holographic reconstruction is obtained when the amplitude transmittance of the hologram is linearly related to the intensity of the interfering object and reference beams. This intensity is given from equation (1.22) by

$$I_T = I_o + I_r + 2\sqrt{I_o I_r} \cos(\phi_o - \phi_r) \tag{6.3}$$

where the holographic grating term is $2\sqrt{I_o I_R} \cos(\phi_o - \phi_R)$. The mean value of I_T is given by

$$\langle I_T \rangle = \langle I_o \rangle + \langle I_r \rangle = \langle I_o \rangle (1 + r) \tag{6.4}$$

where $\langle I_o \rangle$ and $\langle I_r \rangle$ are the mean intensities of the object and reference beams respectively, and $r = \langle I_r \rangle / \langle I_o \rangle$.

Consider now the effect that the value of r has upon the performance of the processed hologram. In all cases it will be assumed that the overall intensity and exposure time are set such that the mean value of exposure energy is at the centre of the linear region of the T/E curve. When r is considerably less than unity the incident light intensity is dominated by the speckle pattern intensity, I_o, and the depth of modulation of the grating term is low. When r is equal to unity the depth of grating modulation will reach a maximum value. Here, however, the resultant swing in the intensity distribution causes the emulsion to operate in the non-linear region. Values of r considerably greater than unity again result in low depths of grating modulation although a high level of linearity can be achieved since the intensity distribution is dominated by the effectively constant reference beam term I_r. Low modulation depth gratings give rise to a dim reconstructed image while non-linearity in the hologram causes noise. It is therefore necessary to compromise between the need for linear performance and depth of grating modulation in order to obtain the best results. Values of r in the range 5 to 10 are generally suitable.

Another requirement for accurate holographic reconstruction is that the emulsion can reproduce all the spatial frequencies in the object/reference interference pattern. The spacing, d, of the grating produced by the interference between object and reference beams is given from equation (1.72) by $d = \lambda/(\sin \theta_o - \sin \theta_r)$ where θ_o and θ_r are the angles of incidence of the object and reference beams at the hologram plane. The value of θ_r is fixed for a plane reference wavefront, but the value of θ_o will vary across the object surface and across the recording plane. The value of d may approach an upper limit of $\frac{1}{2}\lambda$, giving a spatial frequency of $\sim 4000 \text{ mm}^{-1}$, which is within the capability of the highest resolution emulsions.

High resolution emulsions have, however, much lower sensitivities than those with lower resolution, and it may sometimes be necessary to use a lower resolution emulsion if the laser power available is limited: this will result in some loss of definition and brightness in the reconstructed image but holographic fringes of good contrast will still be obtainable.

Typical exposure energies are of the order of 10–50 mJ for emulsions of resolution 3000–5000 mm^{-1}, while emulsions having resolution of $\sim 1000 \text{ mm}^{-1}$ have exposure energies of ~ 1 mJ. The manufacturer's data should be consulted for more accurate values.

Holographic emulsions are often employed as recording media in double exposure SPP and SPC interferometry (Chapter 3) although the lower resolution requirements of these techniques does mean that standard fine grain photographic emulsions such as Pan F may be used. (Care should be taken when using 35 mm roll film to ensure that it is supported against a rigid backing plate in order to prevent movement during exposure.) In both techniques the exposure time should be set such that the transmittance of the developed emulsion is about 0.5. For the double exposure methods this is not too critical but in the case of live SPC interferometry (Section 3.6.2) the process results in fringes of intrinsically low contrast even when the transmittance is of the optimum value of 0.5. Particular attention must therefore be paid to the photographic recording when this form of interferometer is used.

By comparison with photographic emulsion, a television camera operating at a standard scan rate of 1/25 second with a conventional vidicon will generate a 1 V peak to peak output for light intensities of 0.2 μw cm^2 at $\lambda = 633$ nm (see also Table 4.1). Typical signal to noise ratios of 55 dB and maximum resolutions of approximately 50 line mm^{-1} can be obtained. The signal to noise ratio will decrease as the light level decreases and image degradation will occur. This does not happen when a photographic emulsion is used since,. within the limits of reciprocity failure and system stability (Section 6.4.1), the information content of the image may be maintained simply by increasing the exposure time. (An analogous effect can in principle be achieved by decreasing the camera scan rate, but this is difficult and may be accompanied by other limitations such as picture flicker.)

6.3.2 *The intensity of expanded laser beams*

In this section, an expression is derived which relates the intensity of an expanded laser beam to the power of the unexpanded beam.

An unexpanded laser beam is often considered to be a plane wavefront of uniform intensity. A more accurate description must take into account that

(i) the beam is divergent (the angle of divergence is typically $\sim 10^{-3}$ radians); and

(ii) the intensity varies across its width.

For most applications in optics the laser is tuned to operate in the TEM$_{oo}$ mode where the variation of the intensity along any radial direction in a plane normal to the direction of propagation is described by the Gaussian distribution (2)

$$I(r) = I_0 \exp - (r/r_0)^2 \qquad (6.5)$$

where I_0 is the intensity at the centre of the beam, $I(r)$ is the intensity of a point in the plane which is located at a distance r from the centre, and r_0 increases as the beam propagates.

It has been found that an approximate value of r_0 for a given laser beam at a given plane can be found experimentally by measuring d_B, the diameter of the spot produced by the beam on a matt black surface: r_0 is then given by $r_0 \simeq \frac{1}{4} d_B$. (The distribution given by equation (6.5) gives 95% of the intensity falling within $4r_0$). The value of r_0 for a given laser is generally given in the laser handbook. It is typically 1–3 mm, the size increasing as the laser power increases.

I_0 is related to P, the laser output power, as follows. The energy passing through an annulus of radius r, width dr, in unit time is given by

$$dP = \frac{dE}{dt} = 2\pi r \, dr \, I(r) \tag{6.6}$$

so that

$$P = \int dP = 2\pi \int_0^\infty r I(r) \, dr = I_0 \pi r_0^2$$

Hence

$$I_0 = \frac{P}{\pi r_0^2} \tag{6.7}$$

Thus, the intensity at any point in the beam cross-section can be found from the spot size and the output laser power.

When the laser beam is focussed and then expanded by a lens of focal length f, a ray passing through a point located at a distance r from the centre of the beam travels at an angle θ to the optic axis where $\theta \simeq r/f$, and θ is small (see Figure 6.2). At a distance l from the focal point, this ray passes through a point located at a distance R from the centre of the beam where $R = l\theta = lr/f$.

The energy passing through an annulus of radius r, width dr is the same as the energy which passes through an annulus of radius R, width dR. Hence

$$I(R)2\pi R \, dR = I(r)2\pi r \, dr$$

Substituting $r = (f/l)R$ and $dr = (f/l) \, dR$, we have

$$I(R) = I(r)\left(\frac{f}{l}\right)^2$$

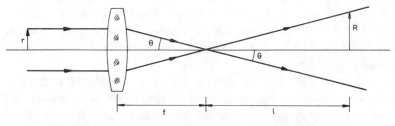

Fig. 6.2 The geometry of a focussed and expanded laser beam

which may be written as

$$I(R) = P \frac{f^2}{\pi r_0^2 l^2} \exp - \left(\frac{fR}{lr_0} \right)^2 \tag{6.8}$$

This expression can be used to find the intensity of an expanded laser beam of known power and beam size.

It can be shown that if it is required to illuminate an object of radius S the maximum intensity at the edge of the object is obtained when

$$\frac{l}{f} = \frac{S}{r_0} \tag{6.9}$$

The intensity at the centre of the expanded beam decreases as $(f/l)^2$. The intensity at a point where $R \neq 0$ depends, however, both on $(f/l)^2$, and $\exp - (fR/lr_0)^2$ and differentiation of equation (6.8) shows that the maximum value of $I(R)$ is obtained when $l/f = R/r_0$.

The normal requirement in holographic and speckle interferometry is that the object illumination is at, or above, a minimum level over the area being inspected, so that generally the object is located at a distance l from the focus of the expanding lens where l satisfies equation (6.9).

In this case the intensity at the edge of the object is given by

$$I(S) = \frac{P}{\pi S^2} e^{-1} = \frac{0.37P}{\pi S^2} \tag{6.10a}$$

and the intensity at the centre is given by

$$I(0) = \frac{P}{\pi S^2} \tag{6.10b}$$

Thus, equation $(6.10a)$ can be used to find the maximum intensity of illumination which can be obtained at the edge of an object of a given diameter with a laser of a given input power.

It can be seen from equations $(6.10a)$ and $(6.10b)$ that the intensity at the edge of the object is only about one-third of the intensity at the centre and if a more uniform object illumination is required, a larger value of l/f must be used. The overall level of illumination will then be reduced.

Finally, it should be noted that equation (6.5) and therefore the rest of the above analysis applies rigorously only to lasers operating just above threshold. It can be used for low power lasers such as HeNe lasers, for argon lasers operating in the lowest part of their power range and for well aligned pulsed lasers.

Microscope objective lenses are frequently used to expand laser beams and it is worth noting that the relationship between the magnification, M, of the lens and its focal length is given by:

$$f \simeq \frac{160}{M}\, \text{mm} \qquad\qquad (6.11)$$

6.3.3 *Light scattering and surface finish*

When an object is illuminated by a light beam, the way in which the intensity of the scattered light varies in space depends on the shape of the illuminating wavefront, on the material of the object as well as on its shape and surface finish. (For a detailed treatment of the scattering of electromagnetic radiation, see Beckmann and Spizzichino (3).) When the illuminating beam is monochromatic, the spatial variation of the intensity of the scattered light is further complicated by the appearance of speckle. In this section, we are concerned with variations in the intensity whose scale is much coarser than the speckle variations, and hence the intensity at a point will be defined here as the mean intensity averaged over many speckles in a region around that point.

At one extreme in scattering surfaces is a mirror finish or specularly reflecting surface. The way in which light is scattered from such a surface can be described by Snell's law (see 1.5.1): each point on the surface reflects an incident ray at an equal and opposite angle to the angle of incidence (see Figure 1.6). The intensity of the reflected light at a given point is determined by the shapes of the surface and the wavefront, and by the object material. A uniform plane wave reflected from a flat surface will have uniform intensity at all points in the reflected wavefront, and the ratio of the incident to reflected intensities may vary between values approaching unity (e.g. gold) to 0.05 or less (e.g. uncoated glass). When a wavefront is reflected from a curved surface, the curvature of the reflected wave is different from that of the incident wave, and the intensity of the reflected wave will decrease as its divergence and its distance from the surface increases.

At the other extreme in scattering surfaces is a diffusely scattering surface, in which each point on the surface scatters the light uniformly in all directions. The intensity of the light scattered from a given point falls off as $1/l^2$ where l is the distance from that point, the intensity of the light scattered from a diffuse surface of area A measured at a distance l may be written as

$$I(l) = \rho \frac{A}{l^2} I_0 \qquad\qquad (6.12)$$

where I_0 is the illuminating intensity and ρ is a constant for a given type of surface. The value of ρ for a surface coated with matt white paint has been measured as $8 \times 10^{-8} \, \text{m}^2 \, \text{mm}^{-2}$ (i.e. ρ is the fraction of the incident intensity scattered by an area of 1 mm^2 to a distance of 1m).

The scattering properties of most surfaces are intermediate between these two extremes. A point on such a surface can be considered to scatter the light into a cone centred on the specular direction. The apex angle χ of this scattering cone increases and the intensity in the specular direction falls off as the magnitude of the fine-scale fluctuations of the surface increases (i.e. as the surface roughness increases). A value for $\rho = \rho_s$ for the specular direction can be determined for a given type of surface by illuminating the surface with an unexpanded laser beam (the area illuminated should be reasonably flat). In this case the maximum intensity $I(l)$ of the scattered light is measured; the mean intensity of the unexpanded laser beam may be taken to be the intensity at the $1/e$ points given by equation (6.7) as $I_0 \simeq 0.4P/A$ where P is the laser power, and A is the laser beam diameter. ρ_s is then given by equations (6.12) as

$$\rho_s \simeq \frac{l^2}{0.4P} I(l) \qquad (6.13)$$

(To compare the measured value with the value quoted here for a diffuse scatterer, l should be measured in metres, and $I(l)$ in power/mm^2). The value of χ can be found by measuring the angle at which the intensity falls to $1/e$ of its maximum value.

A surface-ground steel surface was found to have a value of $\rho_s = 3.1 \times 10^{-6}$ and $\chi = 1°$ in one direction and $6°$ in the other. A surface sprayed with silver paint was found to have $\rho_s = 1.4 \times 10^{-6}$ and $\chi = 18°$.

To find the object beam intensity in the holographic recording plane, the intensity of the light scattered by all points on the surface to a given point on the hologram plate must be found. When the surface is a diffuse scatterer this can be done using equation (6.12), A being the illuminated object area and l the object-to-hologram-plate distance.

When the object has an intermediate type of scattering surface it is much more difficult to estimate the recording plane intensity. It can be seen that, unless the scattering cones of all points on the object intersect the hologram recording area, some parts of the object will not be visible in the reconstruction. In some cases the scattering cones for the whole surface may not intersect at a single plane (see, for example, Figure 6.3), and in some cases, it may be necessary to locate the hologram at a considerable distance from the object, so that the object beam intensity may be very low. In such cases, it is usually more efficient in light usage

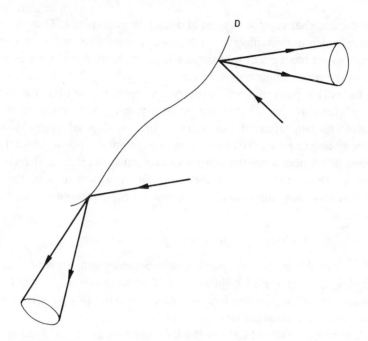

Fig. 6.3 Cones of light reflected from a quasi-specular surface, D

to make the surface diffusely reflecting by coating it with white matt paint or magnesium oxide. (The latter is conveniently deposited by burning a strip of magnesium ribbon approximately 20 cm below the surface to be coated.)

If ρ_s and χ are measured for the surface, these can be used in conjunction with ray tracing to determine whether or not a plane exists where the scattering cones overlap, and if such a plane does exist, the intensity there can be calculated. It can then be seen whether the object should be given a diffusely scattering finish.

If the value of I_0 given by equation (6.7) for a laser of power P illuminating an area πS^2 is used it can be seen that the relationship between the laser power, P, and the minimum holographic recording plane intensity, I_{rec}, is very simple, being given by

$$I_{\text{rec}} = \frac{0.4\rho P}{l^2} \tag{6.14}$$

where l is the object-to-hologram-plane distance. Thus if the object surface is diffuse, the laser power required to achieve a given object beam intensity at the hologram recording plane is determined only by the distance between the object and the hologram plane. Thus the main factor which determines the laser power required to make a hologram of an object having a diffuse surface is the stability of the system, which limits the holographic exposure time.

In speckle pattern interferometry, an image of the object is formed by a viewing lens. The image-plane intensity is determined by the scattering properties of the surface, the viewing geometry and the illuminating intensity. To obtain a uniform intensity image, the scattering cones of all points on the object surface must overlap at the viewing lens aperture. When this is the case, the considerations of the next section show how the image-plane intensity can be found.

6.3.4 *Expressions for image-plane intensities*

In this section the relationship between image-plane intensity, viewing geometry and object-beam illumination is considered. It is assumed that the viewing lens aperture is located to give a reasonably uniform image intensity (see Section 6.3.3).

The intensity of the light at the lens aperture scattered from an area δA of the object is given from equation (6.12) by

$$I_{\text{aper}} = \rho \frac{\delta A}{l_1^2} I_0 \tag{6.15}$$

where l_1 is the object-to-lens distance, I_0 is the object-beam intensity, and ρ is the scattering coefficient.

The energy scattered by the area δA through the aperture in unit time is then given by

$$\delta P = \frac{\delta E}{\delta t} = \pi d_A^2 \cdot I_{\text{aper}} = \frac{\pi a^2}{4 l_1^2} \rho I_0 \delta A \tag{6.16}$$

where a is the diameter of the aperture and δP is the energy which illuminates an area $\delta A'$ in the image plane in unit time; $\delta A'$ is related to δA by

$$\delta A' = m^2 \, \delta A = \frac{l_2^2}{l_1^2} \delta A \tag{6.17}$$

where m is the image magnification and l_2 is the lens-to-image-plane distance.

The intensity in the image plane is then given by

$$I_{\text{im}} = \frac{\delta P}{\delta A'} = \frac{\pi d_A^2}{4 l_2^2} \rho I_0 \tag{6.18}$$

If $l_1 \gg l_2$ is generally the case, then $l_2 \simeq f$ (see equation 1.48) where f is the focal length of the lens. Thus we have

$$I_{im} = \tfrac{1}{4}\pi \frac{\rho I_0}{(\mathrm{NA})^2} \qquad (6.19)$$

where $\mathrm{NA} = (f/d_A)$ is the numerical aperture of the lens. The numerical aperture in speckle photographic experiments is limited by the depth of focus required, or else by the lens itself. It was seen in 4.3.3 that the numerical aperture of the lens in ESPI systems is governed by the spatial resolution of the television camera.

Equation (6.19) can be used in conjunction with equation (6.13) to estimate the image plane intensity which will be obtained for a given laser power, numerical aperture and surface finish.

6.3.5 *Output power and characteristics of laser sources*

In Table 6.2, laser sources commonly used for holographic and speckle measurements are listed with their relevant properties.

6.4 Aspects of mechanical design

When an optical system is set up it is invariably required that beams of light be directed, split and expanded using various combinations of front reflecting mirrors, beamsplitters and lenses. All these components must be mounted in a stable manner and often adjusted with several degrees of freedom. It is not intended to pursue in detail the mechanical design of such mounts except to note that they should be as compact as possible and have a low centre of gravity. We are concerned primarily with the way in which the stability of a system and components can be determined once it has been assembled. This topic is discussed in Section 6.4.1. The fact that many optical components are only used over the diameter of an unexpanded laser beam means that they do not have to be of particularly high quality. For example, a reflecting surface of nominal diameter 50 mm worked to a flatness of 2 wavelengths (a fairly slack optical tolerance) will be flat to better than $\lambda/20$ over the beam aperture. This is more than adequate for most applications. Simple singlet lenses can usually be employed for image formation in SPC interferometers. Low F-number devices such as microscope objectives and collimating lens must, however, be of high quality when they are operated over their full aperture and diffraction-limited performance is required (see Sections 1.5.3 and 1.6.2).

Table 6.2 *Characteristics of laser sources*

Type	Continuous or pulsed	Wavelength	Variable	Power or energy/pulse	Coherence	Power and cooling	General comments
HeNe	Continuous	632.8 nm	No	0.1 mw–50 mw	5–30 cm	Mains; air cooling	Relatively inexpensive and convenient. Used for holographic and speckle where limited power is sufficient
Argon	Continuous	476–514 nm	Several discrete wavelengths in this range	100 mw–15 w	5 cm–several metres (long coherence lengths require use of an etalon)	Generally 3 phase; water cooling	Quite expensive, but generally used in holographic and speckle experiments for larger objects. Also used in two-wavelength shape measurement. (Lower powered, portable, air cooled argon lasers operating at 488 and 514 nm lines only, without etalon are available)
Krypton	Continuous	468–678 nm	Several discrete wavelengths in this range	100 mw–15 w	5 cm–several metres	3 phase; water cooling	Useful if higher power operation in the red region of the spectrum is required, for example in shape measurement

HeCd	Continuous	442–325 nm	Two lines only	5 mw–10 mw	5–30 cm	Mains; air cooling	Used when UV sensitive recording media, e.g. photoresist is used. More expensive than equivalent HeNe
Dye lasers	Continuous and pulsed	300–700 nm	Continuously tunable over several ranges using various dyes	Up to typically 200 mw	1 cm to several metres	Depends on type	Enormous range in price and quality and may require argon laser pump. No real advantage over other lasers except in two-wavelength shape measurement if wide range of contour interval is required
Ruby	Pulsed	654.1 nm	No	1 to 10 J	Few mm up to several metres	Mains; air cooling	Expensive, but are necessary for some dynamic measurements and for measuring relatively unstable structures

A second area that has an important bearing on experiments where objects need to be repositioned to wavelength accuracy (for example live HI (Section 2.1) or component inspection (Section 5.6) is the design of relocation jigs. The basic principles for their design are outlined in Section 6.4.2.

6.4.1 *The assessment of the stability of experimental systems*

With the exception of double-exposure speckle pattern photography (Section 3.2), the techniques described in this book must be operated under conditions of interferometric stability during the exposure time. This means that random disturbances of the various components due to vibration or temperature fluctuations should have amplitudes of considerably less than one wavelength of the light used during the time the recording is made. This imposes quite severe mechanical and environmental stability requirements on the design of the system.

The base on which the optical components are mounted should be very rigid. A common form of interferometer base consists of a slab of cast iron with a machined face for the mounting of the optical components. This has good thermal properties, exhibits excellent long-term dimensional stability and can be used in conjunction with magnetic mounts. The bed must be mounted on anti-vibration mounts in cases where the stability of the laboratory floor and the surrounding building is inadequate. A firmly supported cast-iron base situated in a ground floor laboratory will often provide adequate stability. Magnetic mounting of components which are not bolted onto the bench is desirable, though the use of fast-setting epoxy resin often provides a convenient and interferometrically stable way of temporarily fixing components. It has been found that the flow of cooling water through water-cooled lasers can cause low level vibrations to be transmitted to the interferometer; therefore it is good practice to decouple such lasers from the main optical bench by mounting on a separate support.

It is important to be able to estimate and also to minimize the time over which the system can be considered to be stable. It would be extremely difficult, not to say tedious, to fully test the mechanical stability of a system each time a holographic or speckle recording is to be made. However, the stability of the environment and the bench can often be established by setting up a Michelson interferometer (Section 1.5.5) on the bench using an unexpanded laser beam as shown in Figure 6.4. The interferometer should be set up using optical mounts of established mechanical stability; such mounts are commercially available.

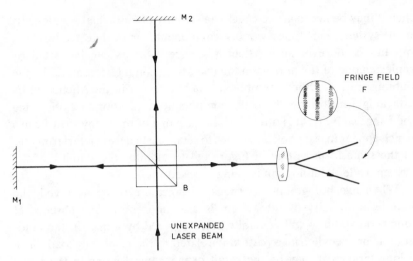

Fig. 6.4 A Michelson interferometer set-up for routine
stability testing

When the interferometer has been aligned the appearance of the
fringe field will indicate the stability of the bed and environment. Stationary fringes indicate an optically stable system. It should be noted however
that high frequency vibrations having an amplitude of less than a
wavelength may not be detected by the eye, and if the presence of such
vibrations is suspected it may be necessary to use a photodetector placed
in the Michelson fringe field to detect the intensity fluctuations.

When the system is unstable the fringes will generally be seen to drift
randomly back and forward across the field of view. This is usually due
either to the transmission of transient ground and air-borne vibrations
or to sporadic thermally induced refractive index variations in the air.
Ground vibrations can be reduced to an insignificant level by mounting
the interferometer on vibration isolation pads. Various types of these
are available commercially and, of these, critically damped air springs
have been found to be effective. Common sense precautions such as
closing laboratory doors and windows, switching off extractor fans, etc.
and then allowing the room to reach a steady state will go a long way
towards reducing air-borne disturbances to an acceptable level. When
the fringe motion has been reduced to a minimum, the maximum allowed
exposure time is determined by estimating the time in which the fringe
drift is less than, or of the order of, one-tenth of a fringe.

Probably the commonest reason for the failure of a hologram is a
lack of mechanical stability. (Other causes of failure are outlined in
Section 6.5.1.) If the procedure outlined above has been followed, and
a very low efficiency holographic reconstruction is nonetheless obtained,

then it may be necessary to check the stability of individual components in the system. Sometimes 'wobbly' components can be detected by hand, but this is more of an art than a science! Mirrors can be tested by replacing one of the mirrors in the test Michelson interferometer by the dubious mirror. Other components may be tested in the Michelson by attaching a mirror rigidly to the component in question and observing the fringe stability. It is often useful to use one of the many stray beams which appear in laser optical set-ups to set up a Michelson interferometer of the same form as that in Figure 6.4 to monitor the stability of the system while the hologram is being exposed.

When live holographic or speckle fringes are to be observed, the long-term stability of the system is also important. The absence of long-term stability will generally be recognized by a drift in the holographic or speckle fringe pattern unrelated to the displacing load. Such a long-term drift may be the result of mechanical creep in the interferometer bed and can be remedied by increasing its stiffness. Alternatively it may be caused by temperature variations; in this case, it can be reduced or eliminated by using materials of low thermal expansion and placing the interferometer in a region of relatively constant temperature.

6.4.2 *The design of relocation jigs based on the kinematic theory of constraints*

In general the problem of reproducibly relocating optical components, hologram plates, inspection components etc. is best solved by adopting a design philosophy compatible with the kinematic theory of constraints. This may be briefly summarized as follows.

A rigid body that is completely free to move possesses six degrees of freedom of motion, three translational and three rotational. If it is desired to contrain the body such that it possesses $n(<6)$ degrees of freedom, then the number of constraints that must be applied is $6-n$. In this context a constraint is defined as being a single point external to the body with which the body is maintained in contact by the application of a forcing mechanism. The latter is often referred to as a 'closure'. The constraints must be applied in such a way that each separately removes one of the undesired degrees of freedom.

As an example let us assume that it is required to constrain a plane surface so that it has two translational degrees of freedom in its plane and a rotational degree of freedom about an axis perpendicular to its plane. The kinematic theory indicates that three constraints need to be applied. These are provided by the three points of support A, B, C

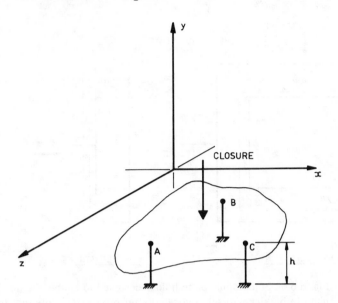

Fig. 6.5 The three constraints, supports at A, B and C, plus closure necessary to kinematically constrain a component to two translational degrees of freedom in the zx-plane together with a rotational degree of freedom about the y-axis

shown in Figure 6.5. A closure acting in a direction perpendicular to the surface must also be applied. (If the object lies in the horizontal plane this closure is provided by the force of gravity). As the height, h, of a support point with respect to a nominally horizontal reference surface is varied it will cause the orientation of the surface normal to change. (This principle is employed in the kinematic design of adjustable mounts.)

Kinematic mounting jigs are designed to enable arbitrarily shaped components to be relocated in a fixed position and orientation. Six constraints corresponding to the removal of the six degrees of freedom are therefore necessary. The actual solution will depend upon the particular nature of the problem but all designs should satisfy the conditions of only six independent constraints. An arrangement suitable for use with a hologram plate is shown in Figure 6.6. (It is advisable to remove any rough edges from the edge of the plate with a fine stone before location in the mount.)

As can be seen from Figures 6.5 and 6.6, point contacts are commonly approximated by the contact of a spherical surface on a plane surface. For example, ball bearings, and if necessary, small plates fixed at

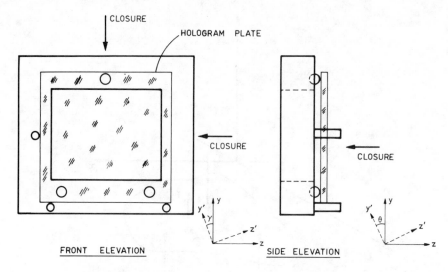

Fig. 6.6 A holographic plate holder designed to enable the hologram plate to be kinematically relocated after processing. Note that the closure forces will be provided by gravity when the holder is rotated by the small angles γ, θ about the z- and x-axis respectively

appropriate positions on the mount and/or component are often used. If the component to be fixed is symmetric in one or more degrees of freedom then the number of constraints is reduced to $6 - n$ where n is the number of degrees of freedom in which the component is symmetric.

6.5 Practical holography

Although holographic systems can often be made quite simple optically it is sometimes the case that the reconstructed images are dim or non-existent. Section 6.5.1 indicates various features of the system that should be investigated when this happens. We then proceed to look at ways of improving the efficiency of holograms (Section 6.5.2) and conclude with a short note on the design of holographic systems with enhanced stability.

6.5.1 *Fault diagnosis of holographic systems*

The considerations of the previous sections on intensity levels, exposure time, processing, stability, kinematic location are all important in obtaining a successful holographic interferogram. If a very low intensity reconstruction (or no reconstruction at all) is obtained, the likely causes of failure in decreasing order of probability are:

(i)   lack of mechanical stability;
(ii) coherence mismatch;
(iii) incorrect exposure, object-to-reference-beam ratio, processing, or the use of plates of incorrect spectral sensitivity;
(iv) stray light;
(v)   mismatch of polarization.

Mechanical stability has already been discussed in Section 6.4.

In the case of (ii) it should be remembered that the interfering object and reference beams must not have a path difference greater than the coherence length of the laser. A piece of string often provides the simplest way of comparing the lengths of the two optical paths.

Exposure levels, reference-beam ratios and processing methods are discussed in Sections 6.3.1 and 6.5.2, if all these appear to be correct, then it is worth replacing the processing chemicals.

Stray light is more likely, in general, to give a lot of noise in the hologram rather than reducing the efficiency of the reconstruction. It should nonetheless be avoided by placing stops at appropriate places in the rig.

If the two beams are orthogonally polarized, no interference will be obtained (see Section 1.3.4). The beams may become orthogonally polarized if one of the beams is rotated about two orthogonal axes when the other is not. A polarizer can be used to determine the polarization of the object and reference beams.

When bright construction of the object is obtained in live HI but either a large number of fringes, or no fringes, are obtained, then the likely causes are:

(i) incorrect location of the plate during exposure or after developing;
(ii) emulsion distortion or shrinkage.

The first of these can be overcome by practice. Techniques for avoiding emulsion shrinkage are discussed in the following section.

6.5.2 *Processing methods for holographic emulsions*

The exposed hologram may be developed, fixed and dried using standard photographic procedures and chemicals. (The same procedure may be used when holographic emulsions are used in double exposure

SPP or SPC interferometry.) Agfa Gevaert recommend the use of their high contrast G3P developer in conjunction with the G334 fixer. Equally good results may be obtained using other forms of developer, for example the fine grain Microphen and Acutol types. Refinements to the process must be made, however, if it is required to either:

> (a) reduce emulsion distortion and shrinkage as, for example, in live holographic or SPC interferometry (Sections 2.3 and 3.6.2); or
> (b) transform the amplitude modulated hologram into a phase hologram by a process of bleaching. This is necessary, for example, when holographic elements for use in two-wavelength shape measurement (Section 5.5.2) are manufactured.

Consider (a). Emulsion distortion can be minimized by soaking the plate in distilled water for about five minutes before use. The plate should be air dried prior to exposure, a process which may be aided by rinsing it in methanol or a proprietary photographic drying agent. This process has the effect of stress relieving the emulsion and should be repeated after the exposed plate has been fixed.

The major advantage gained from bleaching the hologram is a substantial increase in the amount of light diffracted by the hologram into the reconstructed object beam. The improvement in performance is quantified by measuring the diffraction efficiency of a holographic grating manufactured using the bleach process (for example Nassenstein *et al.* (4) pp. 25–38). Bleaching transforms the opaque silver in the amplitude hologram into a transparent silver compound, usually silver bromide. It is the difference between the refractive index of the latter and the emulsion gelatine that creates the required phase hologram. Considerable work has been carried out in this area (4–7) and a process developed by Phillips and Porter (8) appears to give excellent results. As in (a) the hologram is pre-exposure soaked and dried using Drysonal. After exposure it is developed in concentrated Neofin Blue. This developer is hydroquinone based and combines high contrast and fine grain processing characteristics. The nominal development time is five minutes after which the emulsion should appear almost opaque in the exposed region. The plate is then fixed, rinsed and pre-bleach cleansed in a solution of ferric nitrate. It is then placed in the bleaching solution and the bleaching continued for 1 minute after apparent clarity has been reached. Finally it is rinsed in water followed by a drying agent and left to air dry. The constituents of the bleach are as follows:

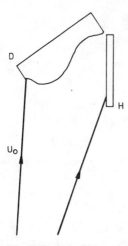

Fig. 6.7 A common path holographic arrangement based on a single object-illumination wavefront

20 g glycerol

500 ml de-ionised water

500 ml isopropyl alcohol

300 mg phenosafranine

150 g ferric nitrate

33 g potassium bromide

This mixture is diluted with four parts of water before use.

6.5.3 *Holographic systems having enhanced stability*

It is required in holographic interferometry that the reference and object beam paths vary with respect to one another as little as possible. Such variation is minimized if the two beams follow common paths or adjacent paths as far as possible. Two 'common path' configurations which are particularly insensitive to external disturbances are shown in Figures 6.7 and 6.8.

Consider Figure 6.7 where a single divergent beam U_o partially illuminates the object D whilst the remainder forms the holographic reference beam. Although this is a simple arrangement it is not particularly versatile.

The addition of the extra reference-beam reflecting mirror, M, shown in Figure 6.8 considerably improves the viewing geometry whilst still

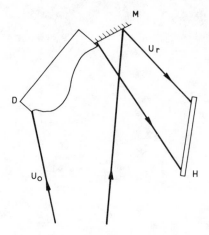

Fig. 6.8 A common path holographic arrangement based on a single object-illumination wavefront and reference beam reflecting mirror

maintaining good stability characteristics. The mirror and the object should be mounted on the same base so that the object and reference beam will move the same distance in the event of the system being disturbed. Rowley (9) has based a holocamera on this arrangement. In this case the illumination source, object and reference beam mirror combination and the hologram recording optics (in the form of a modified single lens reflex camera) are mounted on three separate tripods. Consistently good results have been obtained without recourse to vibration insulation.

In some cases the hologram exposure time needs to be quite long. For example, exposure times in excess of 1 minute are often necessary when UV sensitive photoresist is being used as a recording medium. Under these circumstances it is advaisable to enclose the complete system during the exposure period. This reduces the effects of small random air-borne disturbances which may become cumulatively significant over longer periods.

6.6 Interferometer design and use

When designing an interferometer the main aim should be to minimize the number of optical components and the overall dimensions of the rig. This is relatively straightforward in the case of double exposure SPP in view of the fact that only an object illumination source and a viewing lens are required. Here the main constraints are imposed by the need to minimize fringe pattern error (see Section 3.5.1). This affects the choice of viewing distance and lens type. The accuracy with which the degree of image defocus is measured in the displacement gradient sensitive arrangements is also critical (Section 3.5.1).

SPC and ESPI interferometers are more complex optically and some practical designs have already been outlined in Chapters 3 and 4. In addition to the latter a useful general purpose system is described in Section 6.6.1.

It is also important to be able to assess the magnitude of the displacement under investigation and to control its value. One of the most common reasons for the initial failure of an experiment is that the object motion is incompatible with the sensitivitiy of the interferometer (see Table 6.1). This applies especially in the case of double exposure methods where the experiment is essentially performed 'blind'. In the case of double exposure SPC interferometry the situation is further complicated by the need to form a suitable halo pattern. As has already been noted, however, (Section 3.5.3) this fact can, with care, be exploited and used to measure small strains in the presence of large rigid body motions. The techniques described in Section 6.6.2 enable similar measurements to be made in the presence of rigid body rotations.

6.6.1 *A combined electronic speckle pattern and holographic interferometer*

The interferometric investigation of the deformation and vibration characteristics of a complicated structure is simplified considerably if the fringe pattern is observed in real time. This can be done readily using ESPI. Under certain circumstances it is also necessary to obtain high fringe resolution holographic interferograms of the same event. An arrangement in which both these forms of interferometry are combined is therefore particularly useful. Figure 6.9 shows the layout of such a system. The unexpanded laser source is split into three components by the uncoated, wedge beamsplitter B_1. The transmitted beam ($\approx 90\%$ of the incident intensity) is expanded by the negative lens L_1 to form the object illumination wavefront U_o. (A negative lens is used here so that if necessary the interferometer may be used with a high powered pulse source, Section 2.6.3.) The two reflected beams pass round path compensating loops to form the speckle reference wavefront U_r' and the holographic reference wavefront U_r'' after expansion by the lenses L_2 and L_3 respectively. VA is a variable density attenuator which is used to set the speckle reference beam at the optimum intensity (Section 4.4).

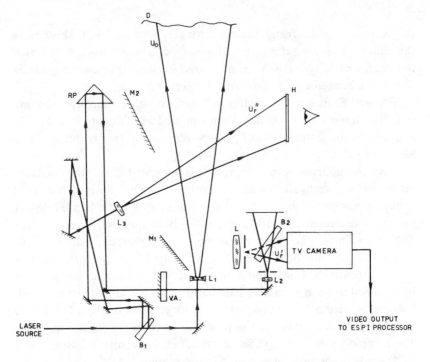

Fig. 6.9 A general purpose electronic speckle pattern and holographic interferometer for deformation and vibration studies

(Neutral density filters can also be used but are less convenient.) Mirrors M_1 and M_2 enable the object, D, to be viewed simultaneously by the ESPI and through the hologram, H. In this way double exposure holograms can be recorded whilst the deformation is being monitored in real time by the ESPI. (When such a system is being used with a pulsed laser it should first be aligned using a CW laser beam aligned to propagate in the same direction as the pulsed source.)

6.6.2 *The elimination of tilt fringes from speckle pattern and holographic fringe patterns*

It is quite often the case in deformation investigations that the body undergoes a localized deformation together with an overall rigid body rotation. The net effect is that the form of the deformation becomes hard to determine and in extreme cases it is completely obscured by the tilt fringes. One obvious solution to this problem is to use a displacement gradient sensitive interferometer (Section 3.7.2) which will differentiate

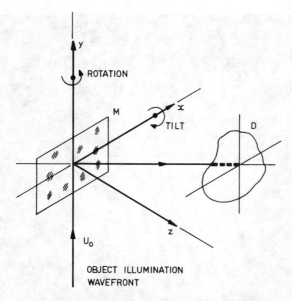

Fig. 6.10 The diagrammatic arrangement of the object illumination geometry for the live elimination of tilt and rotation fringes as used in conjunction with ESPI

the linear displacement gradient to a constant level. An alternative approach is to compensate for the effects of rotation and tilt after the deformation has been applied. Two methods by which this may be done are now described.

Consider the arrangement shown in Figure 6.10 (10). U_o is the on-axis object-illumination wavefront derived from an interferometer of the form shown in Figure 4.10. The light is reflected onto the object, D, by the mirror, M, which is located fairly close to the object and has precision rotation and tilt adjustment. When the object is deformed the tilt fringes may be walked out by applying tilts and rotations to the mirror about the x- and y-axes respectively until a minimum number of fringes are observed on the screen. A typical result is shown in Figures 6.11(a) and (b).

The first interferogram shows a combination of out-of-plane deformation and tilt and in the second the tilt has been removed to reveal the fringes corresponding to the localized deformation. These fringe patterns were obtained as part of the measurement of the displacement distribution of the section of a conformal gear wheel which was subject to static loading. This work is discussed in detail in Section 7.4.1.

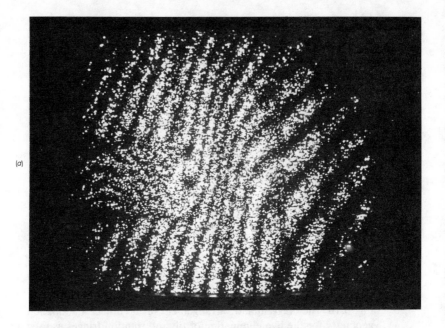

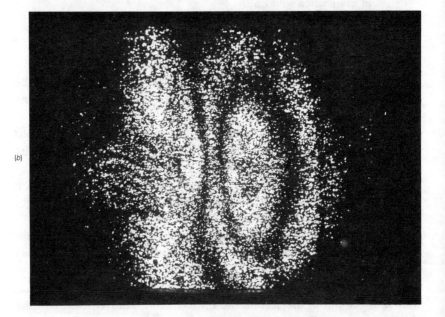

Fig. 6.11 (*a*) ESPI fringes representing a combination of out-of-plane deformation and tilt and rotation. (*b*) The same fringe pattern as that shown in Figure 6.11(*a*) with the tilt and rotation fringes eliminated. The arrangement shown in Figure 6.10 was used

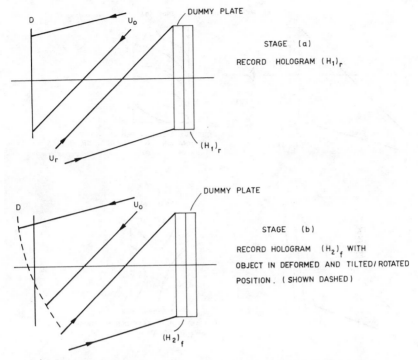

Fig. 6.12 Hologram recording stages in sandwich holography

Consider now a holographic interferometric technique based on sandwich holography (11). An application of this method has already been discussed in Section 5.9.1 where it has been shown how the method simplifies the interpretation of single-wavelength depth contours. The same basic principle may also be used to eliminate tilt fringes from holographic interferograms. Figures 6.12 and 6.13, respectively, show the recording and fringe observation procedures. A hologram $(H_1)_r$ and an identical type of plate, P, with its emulsion removed are placed together in the kinematic plate holder. The emulsion of the hologram plate $(H_1)_r$ is in intimate contact with the dummy plate, P. A hologram of the undeformed object is then recorded by this plate combination (Stage a). These holograms are then removed from the mount and the same procedure is repeated using P and $(H_2)_f$ except that in this case the object has been deformed (Stage b) and the hologram plate is placed in front of the dummy. The processed holograms, $(H_1)_r$, $(H_2)_f$, are then combined to form a sandwich hologram from which the initial holographic fringe field is reconstructed (Figure 6.13, Stage a); this will generally consist of a combination of deformation and rigid body rotation

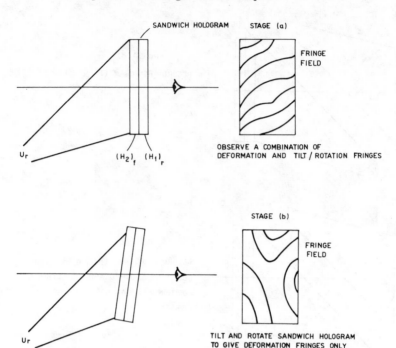

Fig. 6.13 Fringe observation stages in sandwich holography

fringes. However the latter are removed when the sandwich hologram is rotated about its horizontal and vertical axes as shown in Figure 6.13 Stage (*b*). The residual fringe pattern then corresponds to the deformation present. This technique was invented by Abramson who has used it in the investigation of a variety of complex engineering problems. An interesting example (12) is shown in Figures 6.14(*a*) and (*b*). These fringes represent the deflection of a milling machine in which the cutting force has been simulated by static loading. In the first of these (Figure 6.14(*a*)) the sandwich hologram has been tilted to reveal the deformation in the knee of the machine. In the second interferogram (Figure 6.14(*b*)) the hologram tilt has been changed to reveal the deformation in the cutting head.

6.6.3 *The interferometric study of rotating objects*

In certain cases it is important to be able to study the vibration and deformation of components whilst they are rotating at their

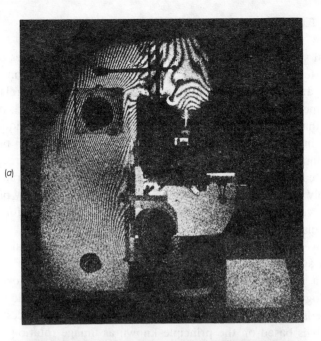

Fig. 6.14 (a) A sandwich holographic study of the deformation of a
milling machine. In this interferogram the sandwich hologram has
been used to eliminate the tilt and rotation of the object in the
region of the knee of the machine. (b) The same basic interferogram
as that shown in (a) with the sandwich hologram adjusted to show
the deformation of the cutting head. (Figure 6.14(a),(b) is
reproduced by kind permission of Dr Nils Abramson)

operational speed. (For example, turbine rotors and disc brakes often need to be tested in this way). Dual pulsed holographic interferometers or speckle pattern photography (Table 6.1) can be used but decorrelation effects (Appendix F) restrict observations to low rotational speeds (typically less than 10^2 rpm). Also, when the first of these techniques is employed, in-plane rotation fringes due to the motion between pulses must be eliminated by observing the object in the specular direction; under this condition the sensitivity to in-plane displacement gradients is effectively zero when the object illumination is approximately plane. (It may be verified that this is a general case of the condition for zero sensitivity discussed in Section 2.5.2).

The above restrictions are eliminated when an opto-mechanical image derotation system is used (13, 14). This arrangement forms a frozen image of a rotating object which may be studied using a focussed image holographic* or speckle pattern interferometer. Considerably higher speeds of rotation may be used and a pulsed laser is not necessarily required. It is based on the principle known as 'image rotation' (15). When a static object is viewed via certain combinations of prismatic optical elements it is found that the image of the object rotates as the element ensemble is rotated about a specific mechanical axis. Such optical devices are known as 'image rotators' and may either be in the form of a transmitting or reflecting element. The configuration of a folded Abbe transmission rotator is shown in Figure 6.15 and the mode of operation of a reflective right angle prism derotator is shown in Figure 6.16. In all cases the angle of image rotation is twice that of the angle of rotation

Fig. 6.15 The folded Abbe image rotator

---

* A focussed image hologram is one in which the holographic reference wavefront interferes with the image of the object formed in the plane of the hologram. Such holograms do not, as a result, reconstruct a three-dimensional virtual image in the conventional manner (Section 1.7). A reconstruction of the two-dimensional image may, however, be observed by illuminating the hologram with white light. The grating structure of the hologram causes the light to be split into its spectrum. An essentially monochromatic reconstructing image may then be seen in the plane of the plate by observing it in the direction of a given spectral order. If a double-exposure or time-averaged hologram is recorded the fringes will be observed on the object image, assuming that they would normally localize in the object plane. (This is the same in principle as the technique described for the white light observation of a double-exposure speckle pattern correlation fringe, Section 3.3).

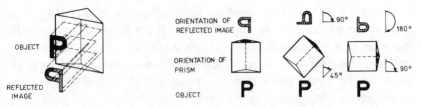

Fig. 6.16 The diagramatic mode of operation of a reflective right-angled prism image rotation system

of the element. It follows that a static image of an object rotating at an angular velocity of ω radians s^{-1} will be obtained provided that it is viewed by means of an image rotation device rotating in the same sense at $\omega/2$ radians s^{-1}. This is the principle of the image derotation system.

Focussed imaged holographic arrangements based on transmission and reflection derotation are shown in Figures 6.17(a) and 6.17(b) respectively. In both arrangements the lens L forms the static image of the rotating object in the plane of the hologram. A folded Abbe element

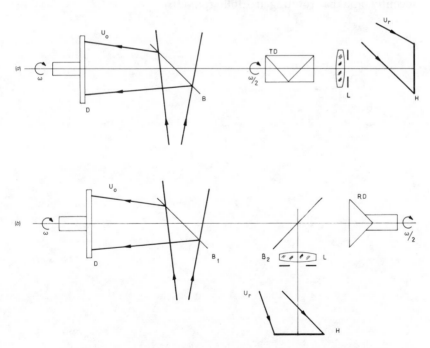

Fig. 6.17 (a) A focussed image holographic system based on a transmission image derotator, TD. (B is a beamsplitter.) (b) A focussed image holographic system based on a reflection image derotator, RD (B$_1$ and B$_2$ are beamsplitters)

(Figure 6.15) is commonly used in the transmission system (14) since this can be readily dynamically balanced. The reflecting element is in the form of a right angle prism (Figure 6.16). This has the advantage that it can be rotated on a central solid spindle since the central aperture need not be transparent; compare Figures 6.17(*a*) and (*b*). Higher rotational speeds are thereby obtained.

For the above cases a focussed image hologram equivalent to that of the rotating object may be recorded using a CW laser. If, for example, the object vibrates during its rotational cycle either as the result of the action of an external source or of a self-generated resonance a time-averaged holographic interferogram of the resonant mode may be recorded. The latter will be observed directly if the output image of the derotator is treated as the object in a smooth reference beam visual speckle pattern interferometer (Section 3.6.4), or ESPI (Section 4.5.2). When the amplitude of vibration is large, dual pulsed holographic interferometry can be used to sample a small portion of the vibration cycle (Section 2.6.3). Results using the last approach have been obtained recently on a disc rotating at 13000 rpm (16).

7

Applications

7.1 Introduction

It would not be practical to describe in a single chapter the numerous phenomena and systems that have been studied using the techniques described in this book. For this reason the contents of this chapter have been arranged so that each section is representative of a general area of application. Each investigation described is relevant to the class of application under consideration. The purpose of this approach is to give the reader some feel for the overall scope of the subject as well as indicating ways in which individual problems may be solved.

7.2 Non-destructive testing (NDT)

The purpose of non-destructive testing is to determine whether or not a component has a fault without causing material damage during the process of the test (see, for example, reference 3). Interferometric methods are suitable in some cases since they are non-contacting and can detect surface displacements that result from the application of very small loads. The main difficulty and skill is in applying the force in such a way that it causes the defect to manifest itself as a discontinuity in surface displacement. The discontinuity appears as an irregularity in the fringe pattern (for example, Figure 7.1) and hence enables the region of fault to be identified. Various methods of creating such a surface displacement may be used, for example:

mechanical loading;
heating;
the application of hydrostatic pressure or vacuum;
vibration.

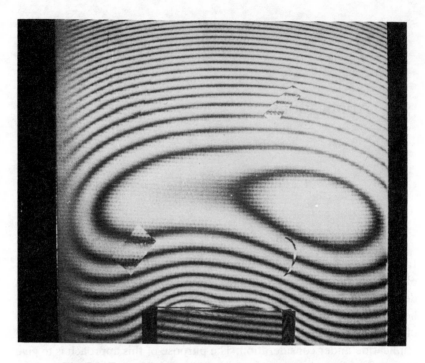

Fig. 7.1 Zones of debond in an epoxy/copper composite as delineated by a holographic interferogram. The deformation was due to slight heating of the specimen. (Reproduced by kind permission of Don Herbert and David Hamson)

Finding the best method and test procedure is partially an empirical process and is a function of the component and the expected nature of the fault. The major limitations of the techniques are:

(*a*) the fundamental requirement that the defect be made to create a discontinuity in the surface displacement; and
(*b*) its poor spatial resolution.

(*a*) means, for example, that voids lying within a stiff matrix cannot be detected easily since it is very difficult to make these affect the surface displacement distribution (such problems are better tackled using ultrasonic (2) or X-ray methods (3)). (*b*) is limited by the minimum resolvable fringe spacing which is typically in the range of 0.5 to 1 mm for viewing areas of about 100 mm diameter (see also Sections 3.5 and 4.3.3). In the case of (*b*) when the size of the defect is small in comparison with the minimum fringe spacing it is clear that it will not affect the overall fringe geometry by a detectable amount. When these limitations are taken into account it is not surprising that the interferometric techniques are most successfully applied to problems where the area of fault is somewhat larger than the minimum fringe spacing for the viewing geometry and exists near the surface of a material which has either relatively low stiffness or is very thin. The detection of debonds in tyres (1) and composite materials – for example, CFRP (carbon fibre reinforced polymer) or metal clad honeycomb structures are common areas of application in which heat or vacuum loading is used to reveal the defect (4, 5). Other areas of investigation include the study of fracture mechanisms (6) and thermally induced stresses (7). Two examples of NDT application are now discussed.

7.2.1 *Debond detection in composite laminates*

The holographic interferogram shown in Figure 7.1 represents the out-of-plane deformation of an epoxy/copper composite sheet caused by heating the copper substrate. Debonds of known shape had been deliberately introduced between the two materials during the process of manufacture. These correspond to the four regions of fringe discontinuity apparent in the pattern. This 'calibration' experiment hence showed that the technique was capable of revealing small zones of debond and could therefore be used for the testing of nominally perfect sheets. In such a test one would subject production composites to the same form of heating, and observe the resultant interferogram. A uniform pattern would be indicative of a uniformly bonded sheet whereas the presence of fringe discontinuities would indicate the presence of debonds. (In this type of measurement it is best to avoid treating the surface under inspection since this can often have the effect of masking the small deformation irregularities associated with the debonds.)

The out-of-plane displacement sensitive electronic speckle pattern interferogram in Figure 7.2 demonstrates the way in which debonds between the honeycomb and the CFRP cladding are revealed under vacuum. Here the region of high density fringes corresponds to the faulted zone. In this test the sample under investigation (approximately 50 mm in width) was placed in a vacuum chamber which was then pumped down. A reference recording (see Section 4.2.1) was made at this stage and a small amount of air admitted to the chamber. The resultant deformation of the surface over the fault is clearly much larger than that in the surrounding unfaulted matrix and enables the extent of the debond to be seen.

Fig. 7.2 Out-of-plane displacement ESPI fringes which indicate a region of debond between a CFPP outer cladding and an internal honeycomb structure. Deformation was produced by a change in the level of vacuum

7.2.2 Thin section weld inspection

In this investigation it was necessary to inspect the quality of strip welds formed between thin ($\simeq 0.1$ mm thick) metallic sections. The electronic speckle pattern interferogram shown in Figure 7.3(a) shows the out-of-plane deformation in the vicinity of a known 'good' weld that resulted from the application of a hydrostatic pressure. This weld had been made under optimum heating conditions. The second interferogram (Figure 7.3(b)) corresponds to the deformation in the vicinity of known poor weld formed under 'cold' conditions. There is a clear difference between the two interferograms. This procedure can therefore be used to identify welds that have been manufactured under different heating conditions and may be employed to test samples from production runs.

7.3 Experimental engineering design investigation

On the basis of the fringe pattern analysis presented in Chapters 2 and 3 it is reasonable to assume that the fringes will generally have a

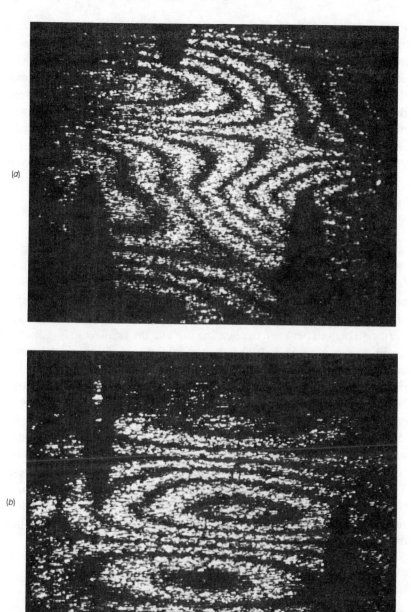

Fig. 7.3 (*a*) The out-of-plane displacement distribution in the vicinity of a known 'good' weld as observed using ESPI. (The deformation was obtained by the application of a hydrostatic pressure). (*b*) As Figure (*a*) but in the vicinity of a known 'cold' weld

higher spatial density in regions of a structure where strain concentrations are present. Furthermore one would expect fringe irregularities to exist in regions of uneven load distribution. Conversely a structure which exhibits uniform load carrying characteristics should generate a deformation fringe pattern of predominantly uniform spacing. It can be seen therefore, that the overall appearance of the fringe pattern may be used both as a qualitative guide to the homogeneity of the deformation of a complex structure and as a means of identifying poor design characteristics. This information may be used in the experimental study of prototype designs and to investigate sources of failure in service.

7.3.1 *A comparative study of engine block deformation characteristics*

The holographic interferograms shown in Figures 7.4(*a*) and (*b*) represent the out-of-plane deformation of the underside and front of the main bearing caps of two V8 engine blocks of different design. (The underside has been illuminated and viewed using a 45° inclined mirror.) The join between the bearing cap and the crank case is indicated by an arrow. A crankshaft has been mounted in position in both cases and loaded statically to simulate the deformation produced by the pistons. In Figure 7.4(*a*), a well-defined discontinuity can be seen between the fringe pattern on the bearing cap and that on the main crankcase. The fringe pattern in Figure 7.4(*b*) is, by comparison, relatively uniform. On the basis of these results one would expect the block in Figure 7.4(*b*) to exhibit the better performance characteristics. This conclusion was supported by the result of a subsequent live engine test. The block in Figure 7.4(*a*) failed in the vicinity of the bearing cap whilst that in Figure 7.4(*b*) maintained good performance throughout the test run. This procedure has been used to study the design of various types of engine block.

7.4 The quantitative measurement of static surface displacements and strain

The quantitative analysis of holographic and speckle fringes presented in Chapters 2 and 3 shows how static surface displacements together with their spatial derivatives (or strain) may be related to fringe pattern geometry and motion. These methods therefore provide a powerful tool for experimental strain analysis, a fact that has been exploited by a considerable number of workers (see for example, reference 8).

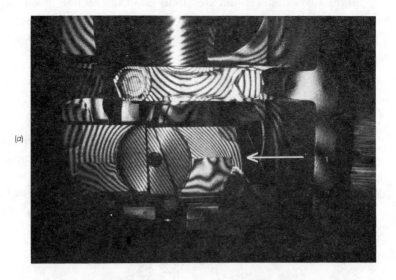

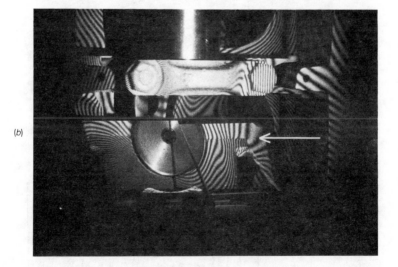

Fig. 7.4 (*a*) A holographic interferogram showing the out-of-plane deflection of a V8 engine block in the region of the main bearing cap due to simulated static piston loading. Note the fringe discontinuity that exists between the bearing cap and the main body of the block. (*b*) An interferogram obtained in the same way as that in Figure (*a*). Note, however, the relative uniformity of the fringe pattern 'throughout the structure

In the two examples that are discussed in this section, it will be seen how the in-plane displacement sensitive configuration of ESPI, holographic interferometry and defocussed SPP have been used to study different problems.

7.4.1 *The verification of a finite element model for the displacement of a conformal gear wheel*

In this investigation the theoretical results for the deflection of a section of a conformal gear wheel as derived from a three-dimensional finite element model were compared with those measured experimentally using ESPI. The experimental loading arrangement is shown in Figure 7.5. The main body of the gear wheel was bolted to a rigid support

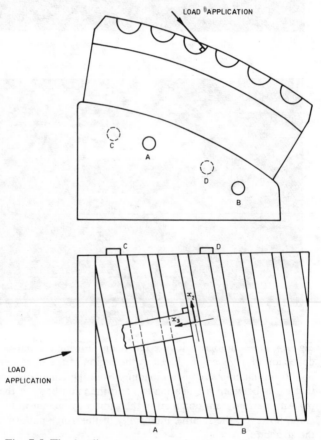

Fig. 7.5 The loading arrangement for the application of deformation to a section of gear tooth

flange at A, B, C and D. An area of tooth of length 10 mm and width 3 mm was loaded in a direction parallel to its surface normal by a hydraulically controlled ram. The region of loading was nominally half-way along the tooth. (This arrangement was not specifically intended to simulate the gear contact load but rather to give a reproducible load input that could also be modelled theoretically.)

The main purpose of the finite element model was to predict the d_2, d_3-deflections in the plane of the tooth as a function of coordinate x_2 (Figure 7.5). In order to do this it was necessary to model the complete wheel section and also take into account the boundary conditions of the support points A, B, C and D. The latter conditions could not be modelled realistically and were therefore determined experimentally using in-plane and out-of-plane displacement sensitive ESPI, i.e. the displacements in the plane parallel to the x_2-, x_3-axes and normal to the plane were measured at A, B, C and D. In all these measurements the magnitude of the applied load was measured using a calibrated load cell and was found to be repeatable. Furthermore the rate of load application was sufficiently well controlled to enable fringe order numbers (and hence displacements) to be determined by counting the number of fringes passing a point in the television monitor image (see also Section 2.4.4). The displacement gradients present were calculated from the geometry of the fringe pattern (for example, Section 3.6.1). When the

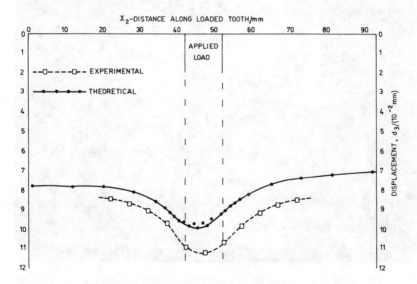

Fig. 7.6 A comparison of theoretical and experimental d_3 tooth-deflection distributions in the vicinity of load application

experimental boundary conditions were incorporated into the theoretical model a d_3-displacement distribution of the form shown in Figure 7.6 was predicted. This displacement profile represents the variation of d_3 with respect to x_2 in the vicinity of loading and is compared with the experimental results. (The latter are plotted as a dashed line.) A typical 'x_3-axis illumination geometry' interferogram (Section 3.7.3) corresponding to the variation of d_3 with respect to the x_2-coordinate is shown in Figure 7.7. These results confirmed that the theoretical model represented a sufficiently accurate description of the physical arrangement to enable it to be used for subsequent theoretical design analysis.

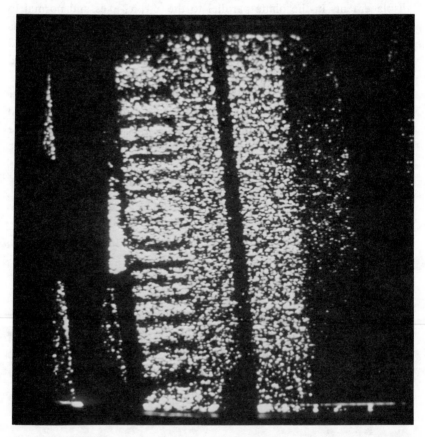

Fig. 7.7 An 'x_3-axis illumination geometry' interferogram that defines contours of constant in-plane d_3 tooth displacement in the region of load application

7.4.2 *The measurement of the dilation of a large hydraulic cylinder*

In this experiment, described by Ennos and Virdee (9), it was necessary to measure the out-of-plane deflection of a hydraulic cylinder of length 1 m and diameter 0.5 m when subject to hydraulic pressure. This measurement was carried out using two independent methods:

(*a*) normal view and illumination holographic interferometry (for example, Section 2.8.1); and
(*b*) defocussed speckle pattern photography (Section 3.3).

The purpose of this approach was to establish the consistency of the results and to verify that the defocussed speckle method had sensitivity comparable to that of holographic interferometry (Section 3.5). The experimental arrangement for the speckle pattern photography is shown in Figure 7.8. Opposite sides of the cylinder were illuminated in a direction effectively coincident with the direction of observation. The plane of defocus of the two viewing lenses was arranged so that the viewing geometry was the same as that shown in Figure 3.8.

Holographic and speckle pattern interferograms corresponding to the same levels of hydraulic pressure were analysed and used to compute the out-of-plane deflection along a cylinder generator. A representative result is shown in Figure 7.9 and indicates that although the holographic and speckle curves have similar shape they give a different absolute value for the deflection. The difference is attributed to a combination of focus error in the speckle pattern photograph (see also Section 3.5) and the fact that tilt was not compensated for in the holographic fringe analysis. More recent investigations (10) in which these errors have been eliminated indicate that the two methods are capable of generating

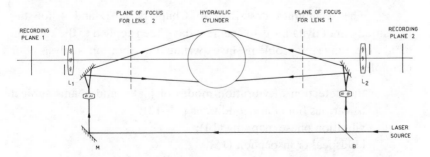

Fig. 7.8 The arrangement for the recording of defocussed speckle pattern photographs of a hydraulic cylinder. (Reproduced by kind permission of Tony Ennos; see Ennos and Virdee (9))

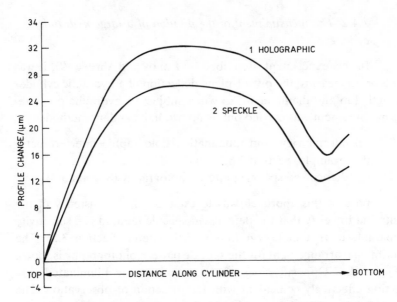

Fig. 7.9 The dilation of the cylinder in Figure 7.8 as measured holographically (1) and by double-exposure speckle pattern photography (2). (Reproduced by kind permission of Tony Ennos; see Ennos and Virdee (9))

identical results. This important result establishes that high resolution measurements may be made using speckle pattern photography without the need for high instrumental stability.

7.5 Experimental vibration analysis

The techniques described in Chapters 2, 3, and 4 for the measurement of dynamic displacement have been applied to the solution of a wide range of problems in experimental vibration analysis. For example:

the detection of vibration modes of plates and beams subject to various boundary conditions (11–13);
vibration phase mapping (14);
loudspeaker inspection (15);
the measurement of two dimensional in-plane vibration modes (16);
the study of transient response vibration modes (17);
the observation of car-body vibrations (18, 19).

Two commonly used methods are time-averaged and dual pulsed holographic interferometry. For this purpose the interferometer shown in Figure 6.9 and described in Section 6.6.1 is particularly suitable. This enables time-averaged or dual pulsed interferograms to be monitored in real time using the ESPI whilst simultaneously recording holographic interferograms using the holographic interferometer. The investigations outlined below were carried out using this system.

7.5.1 *The observation of turbine blade and rotor resonant modes*

If turbine blades and rotors have natural frequencies of response within the range of frequencies that the engine generates as it runs, then fatigue failure of these critical components can occur. Their vibration analysis is therefore an important aspect of jet engine design. It is possible to predict theoretically the engine running frequencies and resonant frequencies of a given blade assembly and this provides the basis for the initial design (20). However, the theoretical analysis is quite involved and it is not possible to reproduce exactly the complex theoretical design profiles and blade orientations (see also Section 7.6.1). For these reasons it is desirable to be able to detect the resonant vibration mode shapes and their frequencies experimentally. The required resonance can be readily detected by the ESPI and a time averaged hologram recorded in the manner described in Section 7.5. For such tests the blade or rotor is conventionally vibrated over the frequency range 1.5 to 50 kHz by sinusoidally driving a piezo electric crystal attached to the surface. (The frequency is monitored using a standard digital frequency counter.) At low frequencies (<1.5 kHz) it is necessary to mechanically excite the component using a directly coupled electromagnetic vibrator.

The above process is straightforward and enables a large number of individual resonances to be recorded in a short period of time. Two typical time-averaged holographic interferometric results are shown in Figures 7.10 and 7.11. Figure 7.10 shows the resonant mode of an individual blade that occurred at 23.15 kHz and in Figure 7.11 a coupled blade and disc resonance is shown. The latter was observed at 12.64 kHz and consists of a combination of diametric disc modes and torsional blade vibration.

7.5.2 *The observation of the vibration modes of a freely suspended electric motor running at high speed*

When turbine blades are tested it is reasonable to clamp the blades rigidly by their fir-tree roots since this is a good approximation

Fig. 7.10 A time-averaged holographic interferogram showing the
resonant vibration mode of an individual turbine blade. The
frequency of resonance is 23.15 kHz

to their operating boundary conditions. Quite often, however, it is
necessary to model theoretically the vibration modes of a complex
structure. This is considerably simplified if in the first instance the effects
of boundary conditions due to, for example, rigid clamping can be
ignored. In practice this requires that the object be freely suspended

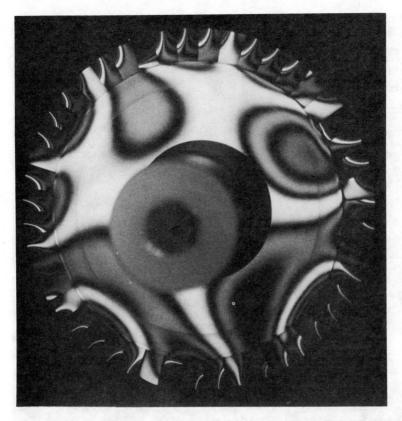

Fig. 7.11 This time-averaged holographic interferogram shows the complex resonant mode of a high pressure turbine disc observed at 12.64 kHz. It consists of distorted diametric nodal regions and coupled torsional blade oscillations.

whilst being tested. Only under these conditions can the experimental results be compared with those predicted theoretically for zero boundary conditions. A problem of this nature involved the measurement of the vibration modes of a high speed motor whilst running at 15 000 rpm. The motor was supported by three rubber bands in order to eliminate boundary conditions and set to run at the required speed. The intrinsic instability of this arrangement ruled out the use of CW methods and dual pulsed holograms of the motor were recorded. A typical result obtained at a 600 μs pulse separation is shown in Figure 7.12. Note that mirrors were placed to one side and below the motor to enable almost a full view to be obtained in a single recording.

Fig. 7.12 A dual pulsed interferogram showing the vibration mode
of a freely suspended electric motor running at 15 000 rpm. A pulse
separation of 600 μs was used. (Experiment carried out by John
Cookson in conjunction with author R.J.)

7.6 Component inspection and quality control

By virtue of the resolution and accuracy attainable optical techniques have provided many useful metrological techniques (21). An
important aspect of metrology is the control and monitoring of the
dimensional tolerance of production components. Various workers have
adapted holographic interferometry so that it may be used to inspect
optical components at wavelength sensitivity (22–25). For many
engineering measurements this sensitivity is too high. The techniques
of two-wavelength ESPI and fringe projection Moiré described in
Chapter 5 provide a sensitivity range compatible with a wide range of

engineering tolerance specifications. Furthermore, they may be used to detect the shape differences between a master and production components of complex geometry. These advantages are exploited fully in the application discussed below.

7.6.1 *Shape comparison techniques applied to the inspection of turbine blades*

This is a problem of considerable commercial importance since improvements in the accuracy of the blades help to reduce the fuel consumption of the engine as well as increasing its overall level of performance. Two parameters have been measured:

(i) the form of the aero-foil section i.e. the accuracy to which the profile of the manufactured blade follows that of the design; and

(ii) the orientation of the aero-foil section relative to a defined axis usually called the stacking axis. Misorientations of the blade with respect to the latter are referred to as 'stacking errors'.

Both measurement techniques require that a blade of known dimension be used as a master for the manufacture of the wavefront projection holograms or Moiré grid negative. (A three-coordinate measuring system may be used to determine the absolute shape of the master. Although this is a fairly time consuming procedure, it need only be carried out once for each design of blade.) At the fringe observation stage (Section 5.6.2) the master blade is removed and replaced by a nominally identical production blade. The resultant fringe pattern will usually consist of a superposition of convoluted profile-error fringes and linear fringes due to a stacking error. This pattern may be analysed using the techniques discussed in Section 5.4.2.

In ESPI the fringe pattern is contained in the output video waveform. (This will also apply in the case of the Moiré technique when the Moiré fringes are observed by a television camera.) If this video waveform is digitized and stored, the position of the fringes may be identified by the computer using suitable fringe-following programs. The purpose of this analysis is to identify the coordinate of the fringe centres and perform the computations necessary to transform them to a plot of shape difference. Figure 7.13 shows dual wavelength ESPI fringes at a contour sensitivity of 25 μm that have been identified by the computer and redisplayed on the television monitor. They correspond to the difference

Fig. 7.13 25 μm sensitivity ESPI contour fringes defining the
difference between a section of master turbine blade and inspection
blade. The position of these fringes has been identified and
enhanced by computer analysis

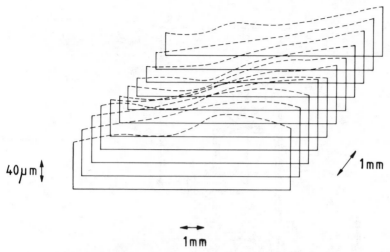

40μm ↕   ╱ 1mm

←→
1mm

Fig. 7.14 An isometric plot of the shape difference between the master blade and inspection blade derived from the interferogram shown in Figure 7.13

between the form of a production blade and master. An isometric plot of this difference is shown in Figure 7.14.

A logical extension of this measurement is to derive a signal from the profile error which will drive a numerically controlled (NC) machine capable of correcting the blade profile. This then establishes a closed loop machining process.

7.7 Fluid flow visualization

The use of holographic interferometry in the observation of refractive index variation of a transparent medium (Section 2.1) provides a particularly useful method of measuring local fluid or gas density variations (26–30). This is because for a wide range of pressures there is a linear relationship between refractive index and density (the Gladstone–Dale relationship, equation 7.6). In the experiment described below (31) these results have been used to measure fuel-vapour-to-air ratios in fuel injection jets.

7.7.1 *The investigation of fuel injection engine combustion processes*

In fuel injection engines fuel is injected into the combustion chamber and a jet-vaporizing of fuel droplets carried by the air entrained

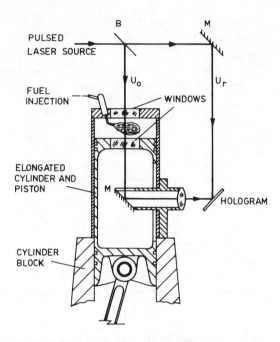

Fig. 7.15 The diagrammatic arrangement of the optical system for the holographic observation of vaporizing fuel droplets in a fuel injection engine

initially in the inspection nozzle is obtained. A knowledge of the local values of fuel-to-air ratios in this spray together with its shape has an important bearing on the combustion efficiency (31) and enables the process to be more fully understood. The experimental arrangement shown in Figure 7.15 is used. The illumination source is a pulsed laser and it will be seen that the optical arrangement is a holographic version of the Mach–Zehnder interferometer (Section 1.5.5). Observations are carried out in two ways. In the first of these the laser is triggered to record a hologram of the combustion chamber at the point of interest in the injection cycle but in the absence of fuel injection. A second recording is then made on the same hologram plate at the same point in the cycle with fuel injection present. When a CW laser is used to reconstruct the double exposure hologram, a fringe pattern revealing the refractive index distribution and hence the density distribution (equation 7.6) in the spray is observed. An interferogram characteristic of the injection of pentane into the cylinder is shown in Figure 7.16. Alternatively one can record a photograph of the spray by triggering

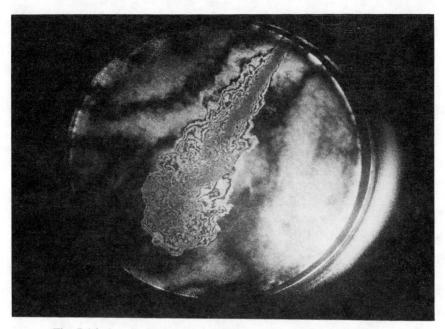

Fig. 7.16 A pulsed laser recorded holographic interferogram
showing the refractive index variation in a fuel injection spray
plume. (Reproduced by kind permission of Professor John Dent)

the single pulse of illumination just after fuel injection has occurred and
observing the shape of the jet from the reconstructed image.

The analysis of interferograms of the type shown in Figure 7.16 is
relatively straightforward if we assume that the jet is axisymmetric about
its direction of propagation. Figure 7.17 shows an infinitesimal element,
$dy\ dz$ of such a jet of coordinate y. (The x-axis corresponds to the
direction of observation and is perpendicular to the plane of the paper
in Figure 7.16.) If the refractive index of this element is $n_0(y)$ in the
absence of the spray and $n_1(y)$ when the spray is present, then the phase
difference between the wavefronts reconstructed from this element is
$d\phi(y)$ where

$$d\phi(y) = \frac{2\pi}{\lambda} [n_1(y) - n_0(y)] \, dy \tag{7.1}$$

The total phase difference, $\phi(y)$, will be given by the integral of the
above equation over the total path length l contained within the jet,
hence

$$\phi(y) = \frac{2\pi}{\lambda} \int_{-\frac{1}{2}}^{+\frac{1}{2}} [n_1(y) - n_0(y)] \, dy \tag{7.2}$$

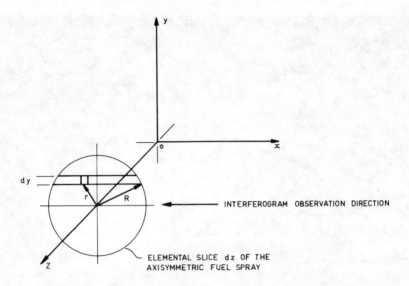

Fig. 7.17 Coordinate system for the analysis of the interferogram shown in Figure 7.16

There will be a maximum of fringe density of fringe order number $N(y)$ when $\phi(y) = 2N(y)\pi$, $N(y) = 0, 1, 2, 3$, therefore

$$N(y) = \frac{1}{\lambda} \int_{-\frac{1}{2}}^{\frac{1}{2}} [n_1(y) - n_0(y)] \, dy \tag{7.3}$$

(The above equation is the same as that used in Mach–Zehnder interferometry for a variation in refractive index perpendicular to the optical path length.) The integral in equation (7.3) may be evaluated if the Cartesian coordinates are transformed to the axisymmetric coordinate system shown in Figure 7.17.

We then find

$$N(y) = \frac{4}{\lambda} \int_{y}^{R} \frac{n_1 - n_0}{(r^2 - y^2)^{\frac{1}{2}}} r \, dr \tag{7.4}$$

where

$n_1 \equiv n_1(r)$ is the refractive index at coordinate r in the presence of the jet;

$n_0 \equiv n_0(r)$ is the refractive index at coordinate r in the absence of the jet; and R is the radius of the jet.

When $n_r = n_0 = 0$ for $r > R$, equation (7.4) may be inverted since it has the form of an Abel integral equation. This yields

$$n_r - n_0 = -\frac{\lambda}{\pi} \int_r^R \frac{[dN(y)/dy]}{(y^2 - r^2)^{\frac{1}{2}}} \, dy \tag{7.5}$$

Experimental values for $dN(y)/dy$ may be measured from the fringe pattern for different values of r and planes of focus of the viewing lens and values of n_r can then be determined from the numerical solution of equation 7.5. n_r is related to the mixture density, ρ_r, by the Gladstone–Dale relationship

$$n_r = \rho_r k_r + 1 \tag{7.6}$$

where k_r is the Gladstone–Dale constant, which, for a binary mixture of air and fuel vapour depends on the mass fraction of fuel vapour and air at r. The use of equations (7.5) and (7.6) therefore enables the local ratios of fuel vapour to air to be calculated.

APPENDIX A

Complex numbers

A complex number z has the form

$$z = x + iy \tag{A.1}$$

where x and y are real numbers and $i = \sqrt{-1}$, so that

$$i^2 = -1 \tag{A.2}$$

x, y are known as the real and imaginary parts of z, and may be written as

$$x = \mathrm{Re}\,(z), \quad y = \mathrm{Im}\,(z)$$

The sum of two complex numbers

$$z_1 = x_1 + iy_1, \qquad z_2 = x_2 + iy_2$$

is

$$z_1 + z_2 = (x_1 + x_2) + i(y_1 + y_2) \tag{A.3}$$

When a complex number z is multiplied by a real number a, we have

$$az = ax + iay \tag{A.4}$$

z may also be written in the form

$$z = r(\cos\theta + i\sin\theta) \tag{A.5}$$

where

$$r = |x^2 + y^2|^{\frac{1}{2}} \tag{A.6a}$$

and

$$\tan\theta = y/x \tag{A.6b}$$

The complex conjugate of z is defined as:

$$z^* = x - iy \tag{A.7}$$

The modulus of z is defined as

$$|z| = |zz^*|^{\frac{1}{2}} = |x^2 + y^2|^{\frac{1}{2}} = r \tag{A.8}$$

It can be shown that

$$\cos\theta = 1 - \frac{\theta^2}{2!} + \frac{\theta^4}{4!} + \cdots \tag{A.9a}$$

and

$$\sin\theta = \theta - \frac{\theta^3}{3!} + \frac{\theta^5}{5!} + \cdots \tag{A.9b}$$

Also, the base of the natural logarithm, e, is defined by the series

$$e^q = 1 + q + \frac{q^2}{2!} + \frac{q^3}{3!} + \cdots \tag{A.10}$$

It follows from equations (A.9a, b) that

$$e^{i\theta} = 1 + i\,\theta + \frac{(i\,\theta)^2}{2!} + \cdots$$

$$= \cos\theta + i\sin\theta \tag{A.11}$$

Hence z may be expressed in the form

$$z = r\,e^{i\theta} \tag{A.12}$$

This is often written in the form

$$z = r\exp i\,\theta \tag{A.13}$$

The product of two complex numbers can then be written as:

$$z_1 z_2 = r_1 r_2 \exp i(\theta_1 + \theta_2) \tag{A.14}$$

APPENDIX B

Vectors

A vector is a quantity which has both direction and magnitude. Examples are velocity, acceleration, electric field. A vector is usually denoted by A, and the magnitude is written as:

$$A = |A| \tag{B.1}$$

A unit vector is a vector whose magnitude is unity, and is often written as $\hat{A}$.

A vector can be represented schematically as a line of given magnitude and direction. Two vectors are equal when their directions and magnitudes are equal.

The sum of two vectors A and B can be represented schematically (see Figure B.1).

A vector may be resolved into component vectors which are parallel to a set of coordinate axes. In Figure B.2 the vector A is seen to be given by

$$A = A_x + A_y + A_z \tag{B.2}$$

If n_x, n_y, n_z are the cosines of the angles formed by A with the x, y, z-axes, we have

$$A_x = n_x A$$
$$A_y = n_y A \tag{B.3}$$
$$A_z = n_z A$$

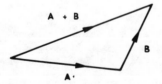

Fig. B.1 The sum of two vectors A and B

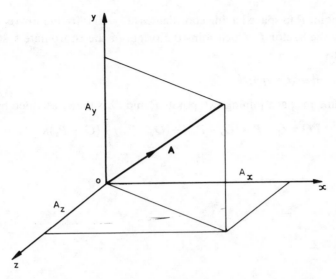

Fig. B.2 The vector **A** resolved into components along the
x-, y- and z-directions

A_x, A_y, A_z are known as the components of **A** in the x-, y- and
z-directions.

If **i**, **j**, **k** are unit vectors in the x-, y- and z-directions, equation (B.3)
may be written as

$$A = A_x i + A_y j + A_z k \tag{B.4}$$

The sum of two vectors **A** and **B** can then be written as

$$A = (A_x + B_x)i + (A_y + B_y)j + (A_z + B_z)k \tag{B.5}$$

The scalar product of two vectors $A \cdot B$ is defined as

$$A \cdot B = AB \cos \theta \tag{B.6}$$

where θ is the angle between **A** and **B**.

By resolving the vectors into their components, it is seen that

$$A \cdot B = A_x B_x + A_y B_y + A_z B_z, \tag{B.7}$$

since

$$i \cdot i = j \cdot j = k \cdot k = 1$$

and

$$i \cdot j = i \cdot k = \cdots = 0$$

A point P in space having coordinates (x, y, z) is frequently represented by the vector $\boldsymbol{P}$ which joins the origin of the coordinate system to P. Thus

$$\boldsymbol{P} = x\boldsymbol{i} + y\boldsymbol{j} + z\boldsymbol{k} \tag{B.8}$$

A line in space joining two points P and Q is then described by

$$\boldsymbol{PQ} = \boldsymbol{Q} - \boldsymbol{P} = (Q_x - P_x)\boldsymbol{i} + (Q_y - P_y)\boldsymbol{j} + (Q_z - P_z)\boldsymbol{k} \tag{B.9}$$

Fourier transforms

A function $g(x)$ which repeats itself periodically at a frequency f can be represented by an infinite sum of sinusoidal waves at multiples of the frequency f as

$$g(x) = \sum_{n=0}^{\infty} a_n \cos (2\pi nfx) + b_n \sin (2\pi nfx)$$

$$\equiv \sum_{n=-\infty}^{+\infty} G_n \exp (i\, 2\pi nfx) \tag{C.1}$$

where a_n, b_n and G_n are constants whose value depends on $g(x)$.

A general function $g(x)$ can be written in the form

$$g(x) = \int_{-\infty}^{+\infty} G(f) \exp (2\pi ifx)\, \mathrm{d}f \tag{C.2}$$

The function $g(x)$ must satisfy certain conditions concerning its smoothness; these conditions are known as the Dirichlet conditions and are given in, for example, references (8)–(10) of Chapter 1. Most of the relationships which describe physical quantities satisfy these conditions.

$G(f)$ is called the *Fourier transform* of $g(x)$; this can be denoted by

$$G(f) \equiv \mathcal{F}[g(x)]; \tag{C.3}$$

$\mathcal{F}$ is the *Fourier transform operator*.

It can be shown that

$$G(f) = \int_{-\infty}^{+\infty} g(x) \exp (-2\pi ifx)\, \mathrm{d}xf; \tag{C.4}$$

This relationship can be written in the form

$$\mathcal{F}^{-1}[G(f)] = g(x) \tag{C.5}$$

where $\mathcal{F}^{-1}$ is known as the *inverse Fourier transform operator*.

It can be seen that

$$\mathscr{F}^{-1}\mathscr{F}[g(x)] = g(x) \tag{C.6}$$

The Fourier transform $\mathscr{F}(g(x, y))$ of a two-dimensional function $g(x, y)$ is found from

$$\mathscr{F}(g(x, y)) = G(f_x, f_y)$$

where

$$g(x, y) = \int\!\!\int\limits_{-\infty}^{+\infty} G(f_x f_y) \exp -2\pi\mathrm{i}\,(f_x x + f_y y)\,\mathrm{d}f_x\,\mathrm{d}f_y \tag{C.7}$$

Some useful properties of Fourier Transforms are as follows:

(i) *Addition*

$$\mathscr{F}^{\pm 1}[g(x) + h(x)] = \mathscr{F}^{\pm 1}[g(x)] + \mathscr{F}^{\pm 1}[h(x)] \tag{C.8}$$

where $\mathscr{F}^{+1} = \mathscr{F}$

(ii) *Multiplication by a constant*

$$\mathscr{F}^{\pm 1}[ag(x)] = a\mathscr{F}^{\pm 1}[g(x)] \tag{C.9}$$

(iii) *Convolution*

$$\mathscr{F}[g(x) \otimes h(x)] = G(f) \cdot H(f) \tag{C.10}$$

where $\otimes$ is the convolution operator and is defined as

$$g(x) \otimes h(x) \equiv \int_{-\infty}^{+\infty} g(x') \cdot h(x - x')\,\mathrm{d}x' \tag{C.11}$$

the integration being performed over all values of x' for a given value of x.

Tilt sensitive double-exposure speckle photography using divergent object-beam illumination

It will be shown first that when the object-beam illumination is divergent (rather than being collimated as in the discussion of 3.3.1), the movement of the speckle pattern in the focal plane of the viewing lens is sensitive to in-plane displacement as well as to in-plane tilt.

The analysis is carried out using coordinate geometry; the origin O of the coordinate system is the point at which the optic axis of the viewing lens intersects the object surface (see Figure D.1), viewing is in the x_1-direction and only displacements in the x_2-directions and tilts about the x_2-axis are considered so that the discussion is confined to two dimensions; the object beam diverges from a point S (s_1, s_2). The viewing lens is located at a distance l_1 from the object surface, and the width of the element illuminating a point in the focal plane is a, where a is approximately equal to the viewing lens aperture diameter. The surface is assumed to lie in the $x_2 x_3$-plane before displacement.

An element centred on the point $P(0, x_2)$ illuminates the point Q in the focal plane of the viewing lens; the line PQ is at an angle α to the

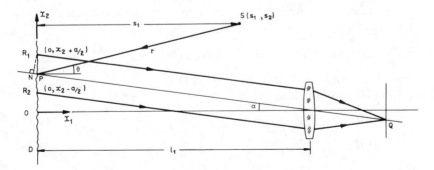

Fig. D.1 Defocussed speckle photography – viewing is in the focal plane, and the illuminating beam is divergent

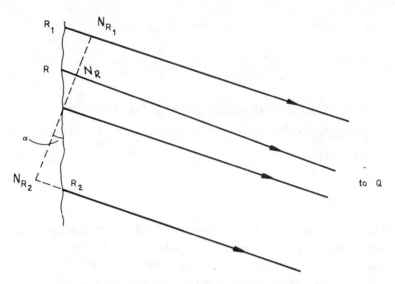

Fig. D.2 The difference in optical path for light scattered from various points in the element centred on P

optic axis where

$$\tan \alpha = \frac{x_2}{l_1} \tag{D.1}$$

The optical path from the point S to the point Q via a general point R in the element is given from equation (1.91) by

$$l_{SRQ} = SR + l_R + RN_R + PQ \tag{D.2}$$

where l_R is the random path associated with the surface roughness at R, and the position of N_R is indicated in Figure D.2.

The difference in optical path from S to Q via the two extreme points $R_1(0, x_2 + \frac{1}{2}a)$ and $R_2(0, x_2 - \frac{1}{2}a)$ is thus given by

$$\Delta l = SR_1 - SR_2 + R_1 N_{R1} - R_2 N_{R2} + l_{R1} - l_{R2} \tag{D.3}$$

where N_{R1} and N_{R2} are shown in Figure D.2.

SR_1 and SR_2 are given by

$$SR_1 = [s_1{}^2 + (s_2 - x_2 - \tfrac{1}{2}a)^2]^{\frac{1}{2}} \tag{D.4a}$$

$$SR_2 = [s_1{}^2 + (s_2 - x_2 + \tfrac{1}{2}a)^2]^{\frac{1}{2}} \tag{D.4b}$$

and

$$R_1 N_{R1} = -R_2 N_{R2} = \tfrac{1}{2}a\alpha \tag{D.4c}$$

When x_2, $a \ll s_1$, s_2 and $r = (s_1{}^2 + s_2{}^2)^{\frac{1}{2}}$, we have

$$(SR_1 - SR_2) \simeq (x_2 - s_2)\frac{a}{r} \tag{D.5}$$

giving

$$\Delta l \simeq \frac{a}{r}(x_2 - s_2) + a\alpha + (l_{R1} - l_{R2}) \tag{D.6}$$

The surface undergoes an out-of-plane rotation $\gamma(\equiv d_{12})$ and an in-plane translation d_2, so that the element A is now centred at $P'(0, x + d_2)$ and the extreme points of A are located at R_1' ($\frac{1}{2}a\gamma$, $x_2 + d_2 + \frac{1}{2}a$), $R_2'(-\frac{1}{2}a\gamma, x_2 + d_2 - \frac{1}{2}a)$. (There will also be a small overall shift in the x_1-direction which can be neglected.) The optical path from S to Q via the various points in the element have clearly altered so that the speckle pattern at Q will have a different amplitude and hence intensity. It will be shown that the relative path lengths scattered from the points in the element to a point Q' which is adjacent to Q are the same as those for the light scattered from the undisplaced element to Q, i.e. the speckle pattern is effectively shifted by the distance QQ'. This may be shown as follows.

The point Q' subtends an angle α' with the optic axis. The difference in optical path from S to Q' via the extreme points Q_1', Q_2' is now given by

$$\Delta l' = SR_1' - SR_2' + 2P'M' + l_{R1} - l_{R2} \tag{D.7}$$

Equation (D.7) is readily evaluated to give

$$\Delta l' \simeq \frac{a}{r}(x_2' - s_2 + d_2 - s_1\gamma) + a(\alpha' - \gamma) + (l_{R1} - l_{R2}). \tag{D.8}$$

When the position of Q' is such that $\Delta l' = \Delta l$, the displaced speckle intensity at Q' is the same as the undisplaced speckle intensity at Q since the relative phases of all the components are unchanged; this is the case when

$$(\alpha' - \alpha) = \gamma\frac{(r + s_1)}{r} - \frac{d_2}{r} = \frac{\gamma}{r}(1 + \cos\theta) - \frac{d_2}{r} \tag{D.9}$$

where θ is the angle between SP and the optic axis. Thus, it is seen that the speckle pattern is rotated by an angle $(\alpha - \alpha')$ whose magnitude is dependent on both the out-of-plane rotation and the in-plane displacement d_2. These two forms of motion are, of course, indistinguishable in the speckle fringe pattern.

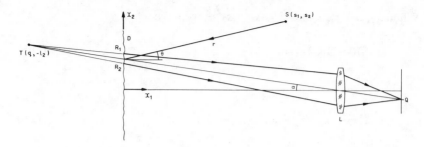

Fig. D.3 Defocussed speckle photography: optical arrangement devised by Gregory (reference 9, chapter 3) to give sensitivity to out-of-plane rotation only using divergent object illumination

It will now be shown that the arrangement devised by Gregory (reference (9), Chapter 3) gives speckle pattern motion which is sensitive to out-of-plane rotation only when a divergent beam is used for the object illumination.

Consider an arrangement (see Figure D.3), where the lens is focussed on a plane situated at a distance l_2 behind the surface. The light arriving at the point Q is scattered from the element A centred on the point P and diverges from a point $T(q_1 - l_2)$ where T, P and Q are colinear. The optical path from S to Q via a general point R in A is given from equation (1.89) by

$$l_{SRQ} = SR + l_R + TQ - RT \tag{D.10}$$

so that the difference in optical path via R_1 and R_2, the extreme points of the element, is given by

$$\Delta l = (SR_1 - SR_2) - (R_1T - R_2T) + (l_{R1} - l_{R2}) \tag{D.11}$$

This can be evaluated to give

$$\Delta l \simeq \frac{a}{r}(x_2 - s_2) - \frac{a}{l_2}(x_2 - q) + (l_{R1} - l_{R2}) \tag{D.12}$$

The object is now translated and rotated as before, and the difference in optical path between the light scattered from the extreme points R_1' and R_2' is

$$\Delta l' \simeq \frac{a}{r}(x_2 - s_2 + d_2 - s_1\gamma) - \frac{a}{l_2}(x_2 - q' + d_2 + l_2\gamma) + l_{R1} - l_{R2} \tag{D.13}$$

where the light now diverges from a point T' $(q' - l_2)$.

The speckle intensity at a point Q' is the same as that at Q when $\Delta l = \Delta l'$; this occurs when

$$-\frac{q}{l_2} = -\frac{(-q' + d_2 + l_2\gamma)}{l_2} + \frac{(d_2 + s_1\gamma)}{r} \tag{D.14}$$

It can be seen that the dependence on d_2 in equation (D.14) vanishes when $l_2 = r$; equation (D.14) then reduces to

$$q' - q = \gamma(r + s_1) \tag{D.15}$$

q and q' are related to the angles α, α', formed by TQ and TQ' respectively with the optic axis, by

$$\alpha = \frac{q}{(l_1 + l_2)}, \qquad \alpha' = \frac{q'}{(l_1 + l_2)} \tag{D.16}$$

so that equation (D.15) can be written as

$$\alpha - \alpha' = \frac{\gamma(r + s_1)}{(l_1 + l_2)} = \frac{\gamma(r + s_1)}{(l_1 + r)} \tag{D.17}$$

Thus the speckle pattern is displaced by a distance δ given by

$$\delta = v\alpha = \frac{v\gamma(r + s_1)}{(r + l_1)} \tag{D.18}$$

where v is the distance between the viewing lens and the recording plane; v is given from the lens equation, equation (1.48), by

$$v = \frac{(l_1 + r)f}{(l_1 + r - f)} \tag{D.19}$$

The speckle pattern is thus displaced by

$$\delta = \frac{(r + s_1)f\gamma}{(l_1 + r - f)} \tag{D.20}$$

which may be written as

$$\delta = \frac{(1 + \cos\theta)fr}{(l_1 + r - f)}\gamma \tag{D.21}$$

Thus, it is seen that when the plane on which the lens is focussed is located at a distance l_2 behind the object, which is equal to the distance of the source S from the object, the movement of the speckle pattern is related only to the out-of-plane rotation of the surface.

APPENDIX E

The correlation coefficient in speckle correlation interferometry

The correlation coefficient of two random variables X and Y is defined as

$$\rho_{XY} = \frac{\langle XY \rangle - \langle X \rangle \langle Y \rangle}{\sigma_X \sigma_Y} \tag{E.1}$$

where

$$\sigma_X = (\langle X^2 \rangle - \langle X \rangle^2)^{\frac{1}{2}}, \qquad \sigma_Y = (\langle Y^2 \rangle - \langle Y \rangle^2)^{\frac{1}{2}}.$$

It can be shown that if X and Y are independent, we have

$$\langle XY \rangle = \langle X \rangle \langle Y \rangle \tag{E.2}$$

so that in this case the correlation coefficient is zero.

The correlation coefficient of $\mathscr{I}_1$ and $\mathscr{I}_2$ defined in equations (3.41) and (3.43) as

$$\mathscr{I}_1 = I_1 + I_2 + 2\sqrt{I_1 I_2} \cos \phi$$

$$\mathscr{I}_2 = I_1 + I_2 + 2\sqrt{I_1 I_2} \cos (\phi + \Delta\phi)$$

can be found by substituting $\mathscr{I}_1$ and $\mathscr{I}_2$ into equation (E.1) giving

$$\rho(\Delta\phi_c) = \frac{\langle \mathscr{I}_1 \mathscr{I}_2 \rangle - \langle \mathscr{I}_1 \rangle \langle \mathscr{I}_2 \rangle}{(\langle \mathscr{I}_1^2 \rangle - \langle \mathscr{I}_1 \rangle^2)^{\frac{1}{2}} (\langle \mathscr{I}_2^2 \rangle - \langle \mathscr{I}_2 \rangle^2)^{\frac{1}{2}}} \tag{E.3}$$

This can be evaluated by noting that

(i) I_1, I_2 and ϕ are independent variables and can thus be averaged separately.

(ii) $\langle \cos \phi \rangle = \langle \cos (\phi + \Delta\phi) \rangle = 0$.

(iii) $\langle I^2 \rangle = 2 \langle I \rangle^2$ (see equation 1.77)

If we assume $\langle I_1 \rangle = \langle I_2 \rangle = \langle I \rangle$, we obtain

$$\rho(\Delta\phi) = \tfrac{1}{2}(1 + \cos \Delta\phi) \tag{E.4}$$

Thus the correlation is unity when $\Delta\phi = 2n\pi$, and zero when $\Delta\phi = (2n+1)\pi$.

If $\langle I_2 \rangle = r \langle I_1 \rangle$, we get

$$\rho(\Delta\phi) = \frac{1 + r^2 + 2r \cos \Delta\phi}{(1+r)^2} \tag{E.5}$$

which has a maximum value of unity when $\Delta\phi = 2n\pi$ and a minimum value of $[(1-r)/(1+r)]^2$ when $\Delta\phi = (2n+1)\pi$.

In this derivation it is assumed that the terms I_1, I_2 and ϕ remain unchanged when the change giving rise to $\Delta\phi$ is introduced. If they are not, then the maximum value of the correlation coefficient is reduced, and if the change is sufficiently large, the speckle patterns will be decorrelated and no fringes will be observed. In Appendix F various factors which give rise to speckle decorrelation are discussed.

APPENDIX F

Speckle pattern decorrelation

In the derivations in this book used to explain the various techniques discussed, it has been assumed that the random amplitude and phase of an individual speckle remains unchanged by the process giving rise to the two states which form the fringes. In holographic and speckle correlation interferometry, the amplitudes in the recording plane before and after recording contain terms of the form

$$U_1 = u_s \exp i\phi_s \tag{F.1a}$$

$$U_2 = u_s' \exp i(\phi_s' + \Delta\phi). \tag{F.1b}$$

and it has been assumed that $u_s = u_s'$ and $\phi_s = \phi_s'$, so that the amplitudes are fully correlated, when $\Delta\phi = 2n\pi$.

In speckle pattern photography, the speckle pattern moves in the recording plane but it is again assumed that $U(r) = U(r')$ where r and r' are the initial and final speckle pattern positions. If

$$U_1 = U(r) = u_s \exp i\phi_s \tag{F.2a}$$

$$U_2 = U(r') = u_s' \exp i\phi_s' \tag{F.2b}$$

it is assumed that $u_s = u_s'$ and $\phi_s = \phi_s'$.

In this Appendix, various factors which cause speckle pattern decorrelation are discussed. For this purpose, it is useful to define the correlation coefficient of U_1 and U_2 as

$$\Gamma(U_1 U_2) = \frac{\langle U_1 U_2 \rangle - \langle U_1 \rangle \langle U_2 \rangle}{\langle U_1^2 \rangle^{\frac{1}{2}} \langle U_2^2 \rangle^{\frac{1}{2}}} \tag{F.3}$$

which can be written as

$$\Gamma(U_1 U_2) = \frac{\langle U_1 U_2 \rangle}{\langle I_s \rangle} \tag{F.4}$$

since it can be assumed that

$$\langle U_1 \rangle = \langle U_2 \rangle = 0 \tag{F.5}$$

and

$$\langle U_1^2 \rangle = \langle U_2^2 \rangle = \langle I_s \rangle \tag{F.6}$$

where $\langle I_s \rangle$ is the mean intensity of the speckle pattern.

In both holographic and speckle correlation interferometry, the visibility of the fringes can be shown to be related to the value of Re $\langle U_1 U_2 \rangle$ which may be written as

$$\text{Re}\langle U_1 U_2 \rangle = \langle I_s \rangle \, \text{Re}\,[\Gamma(U_1 U_2)] \tag{F.7}$$

F.1 Speckle pattern decorrelation due to the variation of the phase, $\Delta\phi$, across an individual speckle

It is assumed when writing equations of the form (F.1*b*) that $\Delta\phi$ is constant across an individual speckle; this is approximately correct when the fringe spacing is much greater than the speckle size. However, as the fringe spacing approaches the speckle size, the approximation is no longer valid and it will be shown that, as the fringe spacing approaches the speckle size, the value of $\Gamma(U_1 U_2)$ is reduced and goes to zero when the fringe spacing is equal to the speckle size.

It will be assumed $\Delta\phi$ varies across a given speckle in one direction only; it is also assumed that the light arriving at a point in the image plane is scattered from a line containing n-point scatterers (this simple model is adequate to give an understanding of this decorrelation process).

The complex amplitude U_1 is given by

$$U_1 = a_0 \sum_{k=-M}^{M} \exp i\psi_k \tag{F.8}$$

where ψ_k is the random phase associated with the kth scatterer, and $2M + 1 = n$.

The phase change giving the second state is introduced, and it is assumed that its value varies linearly across an individual scattering element so that the phase change at the kth point in a given element may be written as

$$\Delta\phi_k = \chi_0 + \tfrac{1}{2}k\,\Delta\chi \tag{F.9}$$

where χ_0 is the value of the phase change at the centre of the scattering element. The amplitude at the image plane is now given by

$$U_2 = a_0 \sum_{k=-M}^{M} \exp i(\psi_k + \chi_0 + \tfrac{1}{2}k\,\Delta\chi) \tag{F.10}$$

The correlation coefficient (equation F.3) is given by

$$\Gamma = \frac{\langle a_0^2 \sum_k \sum_l \exp i[(\psi_k - \psi_l) + \chi_0 + \tfrac{1}{2}k\,\Delta\chi]\rangle}{\langle U_1 U_1\rangle^{\frac{1}{2}} \langle U_2 U_2\rangle^{\frac{1}{2}}} \tag{F.11}$$

and this reduces to

$$\Gamma = \frac{\exp i\chi_0}{n} \sum_{k=-M}^{M} \exp \tfrac{1}{2}ik\,\Delta\chi \tag{F.12}$$

which may be written as

$$\Gamma = Z \exp (i\chi_0) \tag{F.13}$$

where

$$Z = \frac{1}{n} \sum_{k=-M}^{M} \exp \tfrac{1}{2}ik\,\Delta\chi = |Z| \exp i\alpha \tag{F.14}$$

The real part of the decorrelation coefficient is then given by

$$\mathrm{Re}\,(\Gamma) = |Z|(\cos \chi_0 \cos \alpha - \sin \chi_0 \sin \alpha] \tag{F.15}$$

The summation in equation (F.14) is identical to that of equation (1.67) and the magnitude is given by an expression similar to equation (1.68) as

$$|Z| = \frac{1}{n} \left(\frac{\sin \tfrac{1}{2}n\,\Delta\chi}{\sin \tfrac{1}{2}\Delta\chi} \right) \tag{F.16}$$

If N is the number of speckles per fringe, we have

$$n\,\Delta\chi = \frac{2\pi}{N} \tag{F.17}$$

Since generally $n \gg 1$, $\Delta\chi$ is small so that equation (F.18) may be written as

$$|Z| = \left[\frac{\sin(\pi/N)}{(\pi/N)} \right] = \mathrm{sinc}\,(\pi/N) \tag{F.18}$$

When $N = 1$, $\mathrm{Re}\,(\Gamma)$ is zero, and no fringes are then observed in either holographic or speckle correlation interferometry. It has been found that the value of N should be >5 to give reasonably clear fringes in ESPI.

F.2 Speckle pattern decorrelation due to displacement

In holographic and speckle pattern correlation interferometry it is assumed that neither the displacements which give rise to the phase variation causing the fringes, nor other displacements not contributing to this phase variation, significantly alter the random phase and amplitude of the speckle pattern in the fringe observation plane. This is, however, an approximation which is valid only when such displacements are less than some minimum value which depends on the kind of displacement involved and the viewing geometry. For example, in the case of in-plane displacement SPC interferometry (Section 3.7.3) a body may undergo in-plane strain together with an out-of-plane rotation. The effect of the latter will be to reduce the contrast of the plane-strain fringes when the magnitude is sufficient to cause speckle pattern decorrelation. In this section the size of out-of-plane and in-plane translations and out-of-plane and in-plane rotations which totally decorrelate the speckle patterns are indicated. It can be assumed that a displacement or displacement gradient which is less than one-tenth of the given values will not cause decorrelation.

In speckle pattern photography, the object displacement or rotation must be great enough that the speckle pattern in the recording plane moves by at least one speckle if a fringe pattern is to be obtained. It is assumed that the translated speckle pattern remains correlated with the original pattern but this only applies within certain limits which are also given in this section.

F.2.1 *Out-of-plane translation – image-plane viewing*

As the object is displaced along the viewing direction, the relative phase of the components of light scattered from the edges of a resolution element compared with the light scattered from the centre of the element changes; it can be shown that this phase change is 2π when the object moves by an amount Δz given by

$$\Delta z = \sqrt{2u\lambda} \tag{F.19}$$

where u is the object-to-lens distance and λ is the illuminating wavelength. Thus, in this case the decorrelation depends only on u, and the speckle pattern is relatively insensitive to this form of displacement (e.g. if the object-to-lens distance is 500 mm, the speckle pattern will be decorrelated when $\Delta z \simeq 160$ μm).

F.2.2 *In-plane translation – image-plane viewing*

When the body moves in its plane, the speckle pattern in the image plane moves in the opposite direction in proportion to the magnification of the viewing system (this is of course the basis of in-plane sensitive SPP: Section 3.2). If the amount of the movement is small compared with the resolution element size, the change in the speckle pattern at a given point in the image plane is also small. It can be shown that the speckle pattern at a point is decorrelated when the object is translated by an amount Δx equal to q, the resolution element diameter. This may be written as

$$\Delta x = q = \frac{1}{m} (NA)\lambda \tag{F.20}$$

where NA is the numerical aperture of the viewing system given by $NA = f/a$, where a is the diameter of the viewing lens aperture and f is its focal length.

Thus, the larger the magnification of the viewing system, the smaller the in-plane translation which decorrelates the speckle pattern.

As the object is translated in its plane, the speckle pattern is translated; it does not remain identical in form because the light scattered from a given point in the object is incident on the viewing lens at a different angle so that the displaced speckle pattern is not identical in form to the original pattern. When Δx is equal to a, the displaced speckle pattern is totally decorrelated with respect to the original pattern.

F.2.3 *Out-of-plane rotation: image-plane viewing*

When the object is rotated by an angle χ about an axis lying in the object plane, the speckle pattern is decorrelated when χ is given by

$$\chi = \frac{\lambda}{q} \simeq \frac{m}{(NA)} \tag{F.21}$$

In this case, the smaller the magnification, the less the tolerance to out-of-plane rotation.

F.2.4 *Out-of-plane rotation: defocussed case*

In defocussed speckle pattern photography, the speckle pattern in the recording plane is translated when the object is tilted out of its plane; the size of this translation is given by equations (3.30) or (3.34), and it can be seen that in each case the angle at which a given element is viewed is altered; when this angle is sufficiently large, the rays scattered at this angle will no longer be incident on the viewing lens aperture, and in this case the speckle patterns are decorrelated. If the change in angle is $(\alpha - \alpha')$, the object to lens distance is r, and the viewing lens aperture diameter is a, the speckle pattern becomes decorrelated when

$$(\alpha - \alpha')u = a \tag{F.22}$$

F.2.5 *In-plane rotation*

An in-plane rotation of the object gives rise to an equivalent rotation of the speckle pattern. A point at a distance R from the centre of rotation moves by $R\theta$, where θ is the angle of rotation; thus the speckle pattern is decorrelated when

$$\theta \simeq \frac{q}{R} = \frac{\lambda}{R}\frac{1}{m}(\text{NA}) \tag{F.23}$$

F.3 Speckle pattern decorrelation due to wavelength change

The speckle pattern arises from the addition of many components of light having random phases due to the scattering of monochromatic light from a surface whose height varies randomly: it is therefore to be expected that if the wavelength of the light is altered, the phases of the individual components are altered so that the resultant speckle pattern changes.

If the surface height varies with a standard deviation of σ, the mean phase difference $\Delta\psi$ between the scattered components at λ_1 and λ_2 may be written as

$$\Delta\psi = 2\pi\left(\frac{1}{\lambda_1} - \frac{1}{\lambda_2}\right)\sigma \tag{F.24}$$

when the surface is viewed and illuminated normally. Thus, when $\Delta\chi = 2\pi$, the speckle patterns are decorrelated – i.e. when

$$\left(\frac{1}{\lambda_1} - \frac{1}{\lambda_2}\right)^{-1} \simeq \sigma. \tag{F.25}$$

If the illumination is at an angle θ to the surface and viewing is in the specular direction, decorrelation occurs when

$$\left(\frac{1}{\lambda_1} - \frac{1}{\lambda_2}\right)^{-1} \simeq \sigma\cos\theta \tag{F.26}$$

When the surface is viewed in a non-specular direction, the decorrelation also depends on the speckle size; it can be shown that the speckle pattern is decorrelated when

$$\left(\frac{1}{\lambda_1} - \frac{1}{\lambda_2}\right)^{-1} \simeq q(\sin \theta_i - \sin \theta_r) \tag{F.27}$$

where θ_i and θ_r are the angles of incidence and viewing.

It has been found that to obtain good visibility ESPI two-wavelength shape-difference fringes, the wavelength difference must be such that

$$\left(\frac{1}{\lambda_1} - \frac{1}{\lambda_2}\right)^{-1} > 16\sigma \cos \theta. \tag{F.28}$$

(Surface finish is usually measured by stylus instruments which give the CLA (centre-line-average) value. The relation between CLA and σ, the standard deviation of the surface height, depends on the distribution of the height – it is given approximately by $\sigma = 1.2 \times (\text{CLA})$ value.)

References

Chapter 1

(1) M. Born and E. Wolf (1970), *Principles of Optics*, Pergamon Press: Oxford.
(2) F. A. Jenkins and H. E. White (1957), *Fundamentals of Optics*, McGraw Hill Book Co.: New York.
(3) R. S. Longhurst (1968), *Geometrical and Physical Optics*, Longmans: London.
(4) W. T. Welford (1962), *Geometrical Optics*, North Holland Publishing Co.
(5) G. R. Fowles (1975), *Introduction to Modern Optics*, Holf, Rinehart and Winston Inc.: New York.
(6) D. Halliday and R. Resnick (1970), *Fundamentals of Physics*, Chapters 34–5, John Wiley and Sons: New York.
(7) N. J. Frank (1950), *An Introduction to Electricity and Optics*, McGraw Hill Book Co.: New York.
(8) D. C. Champeney (1973), *Fourier Transforms and their Applications*, Academic Press: London
(9) J. W. Goodman (1968), *An Introduction to Fourier Optics*, McGraw Hill Book Co.: New York.
(10) R. C. Jennison (1961), *Fourier Transforms and their Convolutions*, Pergamon Press: New York.
(11) R. Brown (1969), *Lasers (A Survey of their Performance and Applications)*, London Business Books: London.
(12) O. S. Heavens (1964), *Optical Lasers*, Methuen and Co. Ltd.: London.
(13) D. Gabor (1949), *Proc. Roy. Soc.*, A197, 454–87.
(14) D. Gabor (1951), *Proc. Phys. Soc.*, **64**, 449–69.
(15) E. N. Leith and J. Upatnieks (1962), *J. Opt. Soc. Am.*, **52**, 1123–30.
(16) J. W. Goodman (1975), *Laser Speckle and Related Phenomena*, Chapter 2, ed. J. C. Dainty, Springer-Verlag: Berlin.
(17) J. M. Burch and J. H. S. Tokarski (1968), *Optica Acta*, **15**, 101–11.

Chapter 2

(1) R. L. Powell and K. A. Stetson (1965), *J. Opt. Soc. Am.*, **55**, 1593–8.

(2) R. E. Brooks, L. O. Heflinger and R. F. Wuerker (1965), *Appl. Phys. Lett.*, **7**, 248–9.

(3) R. J. Collier, E. T. Doherty and K. S. Pennington (1965), *Appl. Phys. Lett.*, **7**, 223–5.

(4) J. M. Burch, A. E. Ennos and R. J. Wilton (1966), *Nature*, **209**, 1015–16.

(5) B. P. Hildebrand and K. A. Haines (1966), *Appl. Opt.*, **5**, 595–602.

(6) J. W. C. Gates, R. E. N. Hall and I. N. Ross (1972), *Optics and Laser Technology*, **4**, 72–5.

(7) R. Brown (1969), *Lasers* (*A Survey and their Performance and Applications*), 35–44, 60–71, London Business Books: London.

(8) E. R. Robertson and J. M. Harvey (eds) (1970), *The Engineering Uses of Holography*, Proceedings of the Symposium, University of Strathclyde, 1968, Cambridge University Press.

(9) I. S. Sokolnikoff (1956), *Mathematical Theory of Elasticity*, 2nd edition, 5–34, McGraw Hill: New York.

(10) J. F. Nye (1957), *Physical Properties of Crystals*, 2–32, Oxford University Press: Oxford.

(11) Reference (9), 23–4, also reference (10), 94–8.

(12) Reference (9), 100–7.

(13) I. Yamaguchi (1969), *Jap. J. Appl. Phys.*, **8**, 766–70.

(14) R. Jones and D. Bijl (1974), *J. Phys. E: Scientific Instruments*, **7**, 357–8.

(15) E. Archbold and A. E. Ennos, reference (8), 387–99.

(16) M. Born and E. Wolf (1970), *Principles of Optics*, 4th edition, 395–6, Pergamon Press: Oxford.

(17) E. Wolf (ed.) (1980), *Progress in Optics*, Vol. XVII, 3–82.

(18) C. C. Aleksoff (1971), *Appl. Opt.*, **10**, 1329–41.

(19) D. Herbert and S. McKechnie, private communication.

(20) R. K. Erf (1974), *Holographic Non-Destructive Testing*, Academic Press: New York.

(21) F. B. Aleksandrov and A. M. Borch-Brievich (1967), *Soviet Physics*, **12**, 258 (*Zh. Tech. Fiz. 37, No. 2, 360*).
704–22.

(22) A. E. Ennos (1968), *J. Phys. E: Scientific Instruments*, Series 2, **1**, 731–9.

(23) K. A. Stetson (1969), *Optik 2*, Heft **4**, 386–400.

(24) T. Tsuruta, N. Shietone and Y. Itoh (1969), *Optica Acta*, **16**, 723–33.

(25) J. W. C. Gates (1969), *Optics Technology*, **1**, 247–50.

(26) J. Tsujiuchi, N. Takeya and K. Matsuda (1969), *Optica Acta*, **16**, 704–22.

(27) N. Abramson (1969), *Appl. Opt.*, **8**, 1235–90.

(28) W. T. Welford (1969), *Opt. Commun.*, **1**, 123–5.

(29) W. T. Welford (1970), *Opt. Commun.*, **1**, 311–14.

(30) S. Walles (1970), *Optica Acta*, **17**, 899–913.
(31) D. Bijl and R. Jones (1974), *Optica Acta*, **21**, 105–14.
(32) R. Jones (1974), *Optica Acta*, **21**, 257–66.
(33) L. Ek and K. Biedermann (1977), *Appl. Opt.*, **16**, 2535–3542.
(34) J. D. Briers (1976), *Optical and Quantum Electronics*, **8**, 469–501.
(35) N. E. Molin and K. A. Stetson (1969), *J. Phys. E: Scientific Instruments* Series 2, **2**, 609–12.
(36) K. A. Stetson (1971), *J. Opt. Soc. Am.*, **61**, 359–1302.
(37) K. A. Stetson and P. A. Taylor (1972), *J. Phys. E: Scientific Instruments*, **5**, 922–6.
(38) K. A. Stetson (1972), *Appl. Opt.*, **11**, 1725–31.
(39) D. Cutter, reference (42), 133–45.
(40) R. E. Hughes, reference (42), 199–221.
(41) P. C. Gupta and K. Singh (1979), *Appl. Phys.* **6**, 233–40.
(42) J. C. Vienot, J. Bulabois and J. Parteur (eds) (1970), *Applications de l'Holographie* Proceedings of the Symposium, Besançon, 1970, Université de Besançon.
(43) E. R. Robertson (ed.) (1976) *The Engineering Uses of Coherent Optics*, Proceedings of the Symposium, University of Strathclyde, 1975, Cambridge University Press.
(44) C. M. Vest (1979), *Holographic Interferometry*, John Wiley and Sons.

Chapter 3

(1) J. A. Leendertz (1970), *J. Phys. E: Scientific Instruments*, **3**, 214–18.
(2) E. R. Robertson and J. M. Harvey (eds) (1970), *The Engineering Uses of Holography*, Proceedings of the Symposium, University of Strathclyde, 1968, Cambridge University Press.
(3) Reference (2), 237–47.
(4) J. M. Burch and J. M. J. Tokarski (1968), *Optica Acta*, **15**, 101–11.
(5) E. Archbold, J. M. Burch and A. E. Ennos (1970), *Optica Acta*, **17**, 883–98.
(6) J. N. Butters and J. A. Leendertz (1971), *J. Phys. E: Scientific Instruments*, **4**, 1–4.
(7) E. Archbold and A. E. Ennos (1972), *Optica Acta*, **19**, 253–71.
(8) H. J. Tiziani (1972), *Opt. Commun.*, **5**, 271–6.
(9) D. A. Gregory (1976), *Optics and Laser Technology*, **8**, 201–13.
(10) P. C. Champeney (1973), *Fourier Transforms and their Physical Applications*, 216–18, Academic Press: London.
(11) R. C. Jennison (1961), *Fourier Transforms and Convolutions for the Experimentalist*, 20–8, Pergamon Press: Oxford.
(12) Reference (10), 32.
(13) H. J. Tiziani (1971), *Optica Acta*, **18**, 891–902.
(14) E. Archibold, A. E. Ennos and M. S. Virdee (1977), *ASPIE*, Vol. 136, (1st European Congress on Optics Applied to Metrology), 258–264.
(15) A. E. Ennos (1980), *Opt. Commun.*, **33**, 9–12.

(16) R. K. Erf (ed.) (1978), *Speckle Metrology*, Academic Press: New York.
(17) Reference (16), 257–66.
(18) R. Jones and J. A. Leendertz (1974), *J. Phys. E: Scientific Instruments*, **7**, 653–7.
(19) J. A. Leendertz, and J. N. Butters (1973), *J. Phys. E: Scientific Instruments*, **6**, 1107–10.
(20) Y. Y. Hung (1974), *Opt. Commun.*, **11**, 132–5. (*See also reference* (16), 61–3.)
(21) R. Jones and J. A. Leendertz (1974), *J. Phys. E: Scientific Instruments*, **7**, 653–7.
(22) E. Archbold, J. M. Burch, A. E. Ennos and P. A. Taylor (1969), *Nature*, **222**, 263–5.
(23) K. A. Stetson (1970), *Optics and Laser Technology*, **2**, 179–81.
(24) L. Ek and N. E. Molin (1971), *Opt. Commun.*, **2**, 419–24.
(25) Y. Y. Hung, I. M. Daniel and R. E. Rowlands (1975), *Appl. Opt.*, **14**, 618–22.
(26) F. P. Chiang and R. M. Juang (1976), *Appl. Opt.*, **15**, 2199–204.
(27) F. P. Chiang and R. M. Juang (1976). *Optica Acta*, **23**, 997–1009.
(28) L. G. Blows and L. H. Tanner (1974), *J. Phys. E: Scientific Instruments*, **7**, 402–5.
(29) J. M. Burch, A. E. Ennos and G. B. Quinn (1980), Presented at Symposium in Memoriam of Dennis Gabor, Israel. To be published in the *Israel Journal of Technology*.
(30) J. M. Burch and C. Farrow (1975), *Optical Engineering*, **14**, 178–85.
(31) C. Farrow (1978), *Optics and Laser Technology*, **10**, 217–21.
(32) J. C. Dainty (ed.) (1975), *Laser speckle and related phenomena. Topics in Applied Physics Vol. 9*, Springer-Verlag: Berlin.
(33) D. A. Chambless and J. A. Broadway (1979), *Experimental Mechanics*, **19**, 286–90.
(34) E. H. Kauffman, A. E. Ennos, B. Gale and D. J. Pugh (1980) *J. Phys. E: Scientific Instruments*, **13**, 519–89.
(35) M. Francon (1979) *Laser Speckle and Applications in Optics*, Academic Press: New York.

Chapter 4

(1) J. N. Butters and J. A. Leendertz (1971), *Journal of Measurement and Control*, **4**, 344–50.
(2) K. Biedermann and L. Ek (1975), *J. Phys. E: Scientific Instruments*, **8**, 571–6.
(3) O. J. Løkberg and K. Høgmoen (1976), *Appl. Opt.*, **15**, 2701–4.
(4) K. Høgmoen and O. J. Løkberg (1977), *Appl. Opt.*, **16**, 1869–75.
(5) K. J. Bohlman (1978), *Closed Circuit Television for Technicians*, Vol. 1, Norman Price Ltd.: London.
(6) G. A. Slettemoen (1977), *Opt. Commun.*, **23**, 213–16.
(7) K. Høgmoen and H. M. Pedersen (1977), *J. Opt. Soc. Am.*, **67**, 1578–83.

(8) R. G. Hughes, reference (43), Chapter 2, 199–218.
(9) T. J. Cookson, J. N. Butters and H. C. Pollard (1978), *Optics and Laser Technology*, **10**, 119–28.
(10) G. A. Slettemoen (1980), *Appl. Opt.*, **19**, 616–23.
(11) D. Herbert, private communication.
(12) G. A. Slettemoen (1979), *Optica Acta*, **26**, 313–29.
(13) R. Jones and C. Wykes (1981), *Optica Acta*, **28**, 949–72.
(14) C. Wykes, J. N. Butters and R. Jones (1981), *Appl. Opt.*, **20**, A50–1.
(15) R. Jones (1976), *Optics and Laser Technology*, **8**, 215–19.
(16) R. Jones and C. Wykes (1977), *Optica Acta*, **24**, 533–50.
(17) R. Jones and J. N. Butters (1975), *J. Phys. E: Scientific Instruments*, **8**, 231–4.
(18) C. Wykes (1977), *Optica Acta*, **24**, 517–50.
(19) R. Jones and C. Wykes (1978), *Optica Acta*, **25**, 449–72.
(20) R. K. Erf (ed.) (1978), *Speckle Metrology*, Academic Press: New York.

Chapter 5

(1) Nelex (1976), *Proceedings of the Conference on Metrology* (NEL East Kilbride, Glasgow).
(2) K. A. Haines and B. P. Hildebrand (1965), *Phys. Lett.*, **19**, 10–11.
(3) S. H. Rowe and W. T. Welford (1967), *Nature*, **216**, 786–7.
(4) N. Abramson (1976), *Appl. Opt.*, **15**, 200–5.
(5) K. Brooks and M. Heflinger (1969), *Appl. Opt.* **8**, 935–46.
(6) B. P. Hildebrand and K. A. Haines (1966), *Phys. Lett.*, **21**, 422–3.
(7) B. P. Hildebrand and K. A. Haines (1967), *J. Opt. Soc. Am.*, **57**, 155–62.
(8) T. Tsuruta, N. Shiotake, J. Tsujiuchi and K. Matsuda (1967), *Jap. J. Appl. Phys.* **6**, 661–2.
(9) J. S. Zelenka and J. R. Varner (1968), *Appl. Opt.*, **7**, 2107–10.
(10) J. S. Zelenka and J. R. Varner (1969), *Appl. Opt.*, **8**, 1431–4.
(11) L. O. Heflinger and R. F. Wuerker (1969), *Appl. Phys. Lett.*, **15**, 28–30.
(12) J. R. Varner (1971), *Appl. Opt.*, **10**, 212–3.
(13) E. S. Marrone and W. B. Ribbens (1975), *Appl. Opt.*, **14**, 23–4.
(14) A. A. Friesem, U. Levy and Y. Silberberg (1976), *The Engineering Uses of Coherent Optics*, ed. E. R. Robertson, Cambridge University Press, 353–73.
(15) M. Born and E. Wolf (1970), *Principles of Optics*, 4th edition, 441, Pergamon Press: Oxford.
(16) J. N. Butters and J. A. Leendertz (1974), *Proceedings of the Technical Program, Electro-Optics Conference (Kiver Communications)*, 43–9.
(17) R. Jones and C. Wykes (1978), *Optica Acta*, **25**, 449–72.
(18) R. S. Longhurst (1968), *Geometrical and Physical Optics*, Section 8.12, Longmans: London.

(19) R. Jones and J. N. Butters (1975), *J. Phys. E: Scientific Instruments,* **8**, 231–4.

Chapter 6

(1) P. Boone and R. Verbient (1964), *Optica Acta,* **16**, 555–67.
(2) D. J. Innes and A. L. Bloom (1966), *Spectra Physics Laser Technical Bulletin,* No. 5 (Published by Spectra Physics Inc.: Palo Alto, California).
(3) P. Beckmann and A. Spizzichino (1963), *The Scattering of Electromagnetic Waves from a Rough Surface,* Pergamon Press: Oxford.
(4) E. R. Robertson and J. M. Harvey (eds) (1970), *The Engineering Uses of Holography,* Proceedings of the Strathclyde Symposium, 1968, Cambridge University Press.
(5) K. S. Pennington and J. S. Harper (1970), *Appl. Opt.,* 1963.
(6) R. L. van Renesse and F. A. J. Bouts (1973), *Optik* (Stuttgart), **38**, 156–68.
(7) A. Graube (1974), *Appl. Opt.,* **13**, 2942.
(8) N. J. Phillips and D. Porter (1976), *J. Phys. E: Scientific Instruments,* **9**, 631–40.
(9) D. M. Rowley (1979), *J. Phys. E: Scientific Instruments,* **12**, 971–5.
(10) R. Jones and C. Wykes (1981), *Optica Acta,* **28** (7), 944–72.
(11) N. Abramson (1975), *Appl. Opt.,* **14**, 981–8.
(12) N. Abramson (1977), *Appl. Opt.,* **16**, 2521–31.
(13) P. Waddell (1975), Proceedings of the 4th Annual Symposium on Incremental Motion and Control Systems and Devices, held April 1–3, University of Illinois, at Urbana-Champaign.
(14) K. A. Stetson (1978), *Experimental Mechanics,* **18**, 67–73.
(15) D. W. Swift (1972), *Optics and Laser Technology,* August, 175–83.
(16) W. F. Fagan, M. A. Beeck and M. Kreitlow (1981), *Optics and Lasers in Engineering,* **2**, 21–32.

Chapter 7

(1) M. J. Cannozaro and F. W. Hill (Jr) (1974), Society of Automotive Engineers, Report of the Automotive Engineering Congress, Detroit, Michigan, Feb 25–March 1.
(2) E. E. Aldridge (1971), *Materials and Research Standards (USA),* **12**, 13–22.
(3) Shi-Tien Li and W. I. McGonnagle (1972), *Int. J. NDT,* **4**, 214–29.
(4) P. J. Kisatshy and M. Barbarisi, *ibid.*
(5) M. J. Marchant (1974), RAE Technical Report No 73192.
(6) T. D. Dudderas and R. O'Regan (1972), *Int. J. NDT,* **4**, 119–47.
(7) W. J. Harris and D. C. Wood (1979), *Materials Evaluations,* **32**, 50–6.

(8) Contributions by various authors: *Journal of Strain Analysis* (1974), **9**, No 1.

(9) A. E. Ennos and M. S. Virdee (1979), Presented at the IUTAM Symposium on 'Optical Methods in the Mechanics of Solids', Poitiens France. To be published in the proceedings.

(10) A. E. Ennos: private communication.

(11) T. R. Leuner (1974), *Journal of Sound and Vibration*, **32** (4), 481–90.

(12) R. Aprahamian and D. D. Evensen (1970), *J. Appl. Mech.*, June, 287–91.

(13) R. A. Aprahamian, D. A. Evensen and J. S. Mixson (1970), *Experimental Mechanics*, October, 1–6.

(14) J. A. Levitt and K. A. Stetson (1970), *Appl. Opt.*, **15**, 145–9.

(15) T. Koizumi (1973), *Mitsubishi Denki Laboratory Report*, **14** (6), 75–90.

(16) E. Archbold and A. E. Ennos (1975), *Optics and Laser Technology*, February, 17–21.

(17) R. A. Aprahamian, D. A. Evensen, J. S. Mixson, J. G. Lacoby and J. E. Wright (1971), *Experimental Mechanics*, July, 1–6.

(18) A. Felske and A. Wolfshung (1973), *ATZ Autobiltechnishe*, **75** (3), 96–102.

(19) A. Fleske and A. Happe (1976), '*The Engineering Uses of Coherent Optics*' pp 595–615. (Proceedings of the Conference held at Strathclyde April 75, edited by E. R. Robertson). Cambridge University Press.

(20) D. J. Ewins (1973), *J. Mech. Engin. Sci.*, **15**, 165–86.

(21) *op. cit.* (19) pp 1–15.

(22) J. M. Burch, C. Forno and L. H. Tanner (1974), *Optics and Laser Technology*, 109–13.

(23) R. W. Evans, P. Gallagher and D. A. Rimmer (1975), *Optics and Laser Technology*, October, 203–8.

(24) N. P. Larirov, A. V. Lukin and K. S. Mustafin (1972), *Optical Technology*,. **39** (3), 154–5.

(25) D. F. Horne and F. E. R. Cannings (1977), *The Production Engineer*, October, 22–5.

(26) R. C. Jayota and D. J. Collins (1972), *J. Appl. Mech.*, December, 897–903.

(27) G. B. Brandt, P. F. Rozelle and B. R. Patel (1975), *Research/Development*, December 897–903.

(28) K. A. R. Kinloch and S. M. Frazer (1976), *J. Phys. D: Applied Physics*, **9**, 1831–8.

(29) A. H. Guenther, W. K. Pendleton, C. Smith, C. H. Skeen and S. Zuri (1973), *Optics and Laser Technology*, February, 29–36.

(30) B. Ineichen, U. Kogelschutz and R. Dunliker (1973), *Appl. Opt.*, **12**, 2554.

(31) J. C. Dent (1980), *Combustion Modelling in Reciprocating Engines*, Eds J. N. Hattavi and C. A. Amann, 265–90, Plenum Press: New York; London.

INDEX